MOLECULAR
BIOLOGY
INTELLIGENCE
UNIT

KINETIC MODELING OF GENE EXPRESSION

James L. Hargrove, Ph.D.
The University of Georgia
Department of Foods and Nutrition
Athens, Georgia, U.S.A.

R.G. LANDES COMPANY
AUSTIN

MOLECULAR BIOLOGY INTELLIGENCE UNIT

KINETIC MODELING OF GENE EXPRESSION

R.G. LANDES COMPANY
Austin

CRC Press is the exclusive worldwide distributor of publications
of the Molecular Biology Intelligence Unit.
CRC Press, 2000 Corporate Blvd., NW, Boca Raton, FL 33431. Phone: 407/994-0555.

Submitted: July 1994
Published: September 1994

Please address all inquiries to the Publisher:
R.G. Landes Company, 909 Pine Street, Georgetown, TX 78626
or
P.O. Box 4858, Austin, TX 78765

Phone: 512/ 863 7762; FAX: 512/ 863 0081

ISBN 1-57059-201-2
CATALOG # LN9201

Library of Congress Cataloging-in-Publication Data

Hargrove, James L.
Kinetic modeling of gene expression / James L. Hargrove.
p. cm.–(Molecular biology intelligence unit)
Includes bibliographical references and index.
ISBN 1057059-201-2
1. Gene expression–Computer simulation. I. Title. II. Series.
QH450.H37 1994
574.87'322'0113416—dc20. 94-32978 CIP

To Our Readers

R.G. Landes Company publishes four book series: *Medical Intelligence Unit, Molecular Biology Intelligence Unit, Neuroscience Intelligence Unit,* and *Biotechnology Intelligence Unit.* Our goal is to publish the most recent information in biomedical science for sophisticated researchers and physicians.

To achieve this goal we have accelerated our publishing program to conform to the fast pace in which information grows in biomedical science. Most of our books are published within *90 to 120 days of receipt of the manuscript.*

As you might expect, this causes a few problems for us; sometimes it makes our office more like a big city newspaper than a scholarly publisher. So as you look through this book you may see something that isn't just right. Please let us know. Or if you have an idea for improving our books, we'd like very much to hear from you. If the problem you describe hasn't already been discovered or if your idea is provocative enough to promote a discussion here, we'll give you a free book of your choice. Just list three titles in order of preference and we'll send you one, based on availability—with our thanks. Our address is printed on the copyright page of each of our books.

Carol Harwell
Publications Director
R.G. Landes Company

CONTENTS

PREFACE

Altered gene expression contributes to the pathology and temporal development of all chronic human diseases, yet there remains a challenge of understanding how altered gene expression contributes to specific physical or metabolic traits. The aim of this volume is to explain the use of microcomputer software as a tool for exploring the dynamic behavior of genetic systems. The presentation attempts to demonstrate linkages between levels in an organism's hierarchy, from the molecular level upwards to cells, tissues, and organ systems. It deals with the importance of time, complexity, scale, and feedback in a way that is not characteristic of biological or medical education. The method recommended here is to begin with the technique called compartmental analysis, and to employ software that interactively generates equations to model the flow of genetic information. Advanced students with a deeper grasp of mathematical modeling may instead choose to employ analytical solutions to differential equations; a few precedents and new resources are described.

Many readers may remember how their typing skills improved with the advent of microcomputer-based word processors. Suddenly, it was permissible to make mistakes because errors could be corrected prior to printing the final copy. Consider that the computer was not designed only to be a word processor, but rather as a tool for solving equations rapidly. Computer software for simulation allows the investigator to harness this capability as a means of comparing predicted and observed responses in biological systems, and one intention of this volume is to show that systems modeling can be just as liberating to quantitative thinking as word processors have been to our ability to create documents. Systems can be designed, and ideas can be tested, for problems ranging from the expression of genes to metabolism, physiology, and disease. The technique of mathematical modeling rightfully belongs in the armamentarium of the contemporary molecular geneticist.

The reader requires no special knowledge to follow the presentation in the text; neither is a special facility with mathematics required, for new computer software helps the beginner solve these problems. The objectives of this presentation are to explain fundamental concepts about the flow and expression of genetic information, and to enable the reader to begin applying concepts derived from systems theory to areas of personal interest. If a few precautions are taken to utilize simple kinetic rules, these goals can be achieved without the need to involve a professional mathematician, or to solve and program complex differential equations. The exercise will be successful if the reader achieves a sense of having understood these principles all along, and, after having read this volume, clearly understands why these familiar principles apply to genetic regulation.

Acknowledgments

In a small way, this volume is dedicated to Professors Dora and Avram Goldstein, who checked Cheston Berlin and Robert Schimke's math, and explained that their theory of enzyme induction is identical to the plateau principle of pharmacokinetics. The interested reader is referred to the footnote in *Molecular Pharmacology* 1965; 1:149-156, and to p. 314 of *Principles of Drug Action*, New York: Harper and Row, 1969, by Avram Goldstein, Lewis Aronow, and Sumner Kalman. It is also dedicated to all those investigators whose quest to quantify gene expression enables this theory to be checked against experimental results. But more than this, it is dedicated to an understanding of the dimension of biological time, in which our health and disease develop, exist, persist, and prevail.

CHAPTER 1

The Role of Computer Simulation in Molecular Biology

There is a more fundamental form of skepticism which the molecular biologist must be prepared to meet, and that is the question: what is the significance of your molecules?
—B.C. Goodwin, Temporal Organization in Cells

INTRODUCTION

The idea that computer simulation may yield fundamentally new insights into the dynamics of biological systems entered the popular consciousness partly through the descriptions of chaos theory and fractal geometry in James Gleick's best-selling book, *Chaos, Making a New Science.*[1] Although software for computer-assisted simulation (mathematical modeling) converts the personal computer into a tool for exploring ideas about complex systems at low cost and great speed, it is safe to predict that few scientists have gained experience with mathematical modeling. This stems not only from lack of familiarity with using mathematics to solve dynamic problems, but also from a proper insistence that ideas should be proved at the bench or in the field. Even so, the founders of genetics, biochemistry, and molecular biology universally understood that important observations must be interpreted in a quantitative framework. A classic example is Mendel's use of probability to describe the independent inheritance of observable traits in garden peas.[2] During a two-year stay at the University of Vienna, Mendel had been influenced by the physicist, Doppler, to use mathematics in quantifying experimental results. Computer simulation enables biologists to take advantage of a powerful, contemporary method for quantitative thinking.

Systems theory affords access to two fundamental questions: How are individual parts of an organism, cell population, or cell quantitatively interrelated, and what is the time dependence of responses to environmental change? Every interesting question in biology and medicine is fundamentally a problem in systems dynamics, whether the specific question deals with the cell cycle, metabolism, the control of growth, or the spread of human diseases. The revolution in personal computing enables investigators who wish to use quantitative, kinetic models to interpret their data to rely on well established precedents in using any of several popular software packages for computer simulation. The aim of this book is to provide guidance to some of the resources that are available, including new computer programs for mathematical modeling, and examples in the literature that can assist the novice in applying the programs to areas of special interest.

A MODEL OF GENE EXPRESSION BASED ON COMPARTMENTAL ANALYSIS

The flow of genetic information through the steps of transcription, mRNA processing and transport, and protein synthesis and degradation is a familiar example of linked processes that are often diagrammed qualitatively (Fig. 1.1). Each gene product presumably contributes to a specific trait or phenotype, and is influenced by signals and stimuli from a network of other cells, tissues, or the external environment. This diagram can be used as a framework for a quantitative, mathematical model by assigning rates to each connection indicated in the diagram; common ways to represent the rate parameters are with the English letter, k, or with the Greek letter, lambda (λ), with subscripts identifying which rate applies to each specific process. The rates need not be constants, as long as the modeler states how a change in input is related to a change in output, or how a stimulus is related to a response. This relationship can be as simple as the frac-

Fig. 1.1. The flow of genetic information that links genes to phenotypes is depicted with rates of synthesis, transfer or degradation that must be included to generate a dynamic model of a genetic system. Each proportional rate parameter has units of time^{-1}, and indicates the fraction of material in each metabolic pool that is produced or transferred per time interval. A complete, mechanistic model must also include a quantitative relationship that links the quantity of protein to a trait or phenotype, and a means of feedback or environmental control of genetic acitivity.

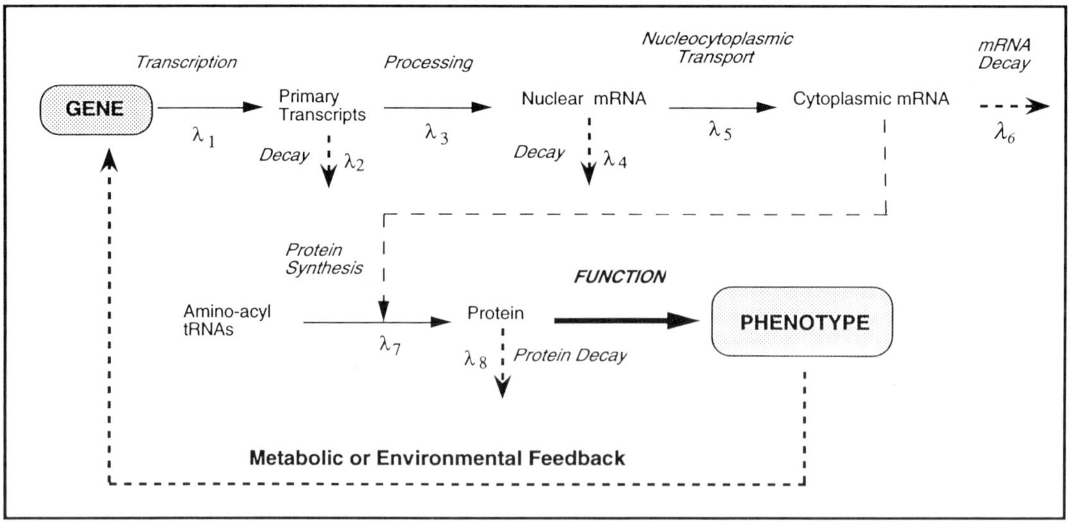

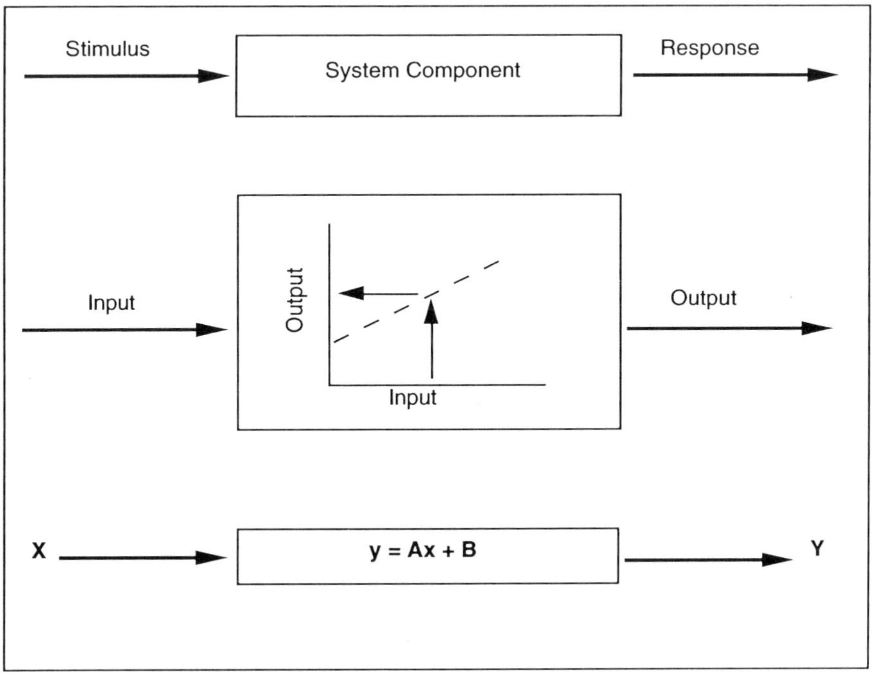

Fig. 1.2. Block diagrams provide a simple means of understanding system dynamics and the modeling process. Each important component in the system is identified, and the relationship between stimulus and response, or input and output, is measured experimentally. The mathematical model is then improved by fitting a curve or algebraic function that describes these relationships. The individual components or blocks are assembled into a master diagram, as shown in Fig. 1.1.

tion of the component of interest that is transferred during each interval, an algebraic equation (Fig. 1.2), or a non-linear saturation function such as the Michaelis-Menten or Hill equation. Investigators usually derive the algebraic relationships from knowledge about the system, and determine the values for the parameters by analyzing experimental data with a curve-fitting routine.

It is notable that the time required to complete transcription, processing, and transport of mRNA is often much shorter than the period needed to achieve a new, steady-state level of protein. The basis for this is not genetic; it is a *physical principle* that predicts rates of accumulation of particles in compartmental systems.[3-6] Cells and organs are compartmental by definition (Old English *cell*, religious house, and Latin *cella*, small room). However, kinetic compartments are not necessarily identical to physical compartments; in the case of gene expression, each intermediate is treated as a separate compartment or pool coexisting within the cell.

Compartmental analysis is a method of quantifying the rates of accumulation, transfer, and elimination of material or information in systems with one or more uniform physical or chemical states.[7-9] It is commonly applied to kinetics of radioisotopes or drugs in systems with multiple compartments, but the unit of function is always a single, identifiable quantity or state (the single compartment) that is linked to other identifiable units. Put simply, the method is block-organized, and in the most basic system, each compartment is characterized by one exponential term that describes the rate at which that compartment approaches a steady state. The method is ideally suited to studies of the flow of genetic information, in which latent information is transcribed, amplified, and translated to generate an enzyme capable of altering the properties of a system. The material of interest is defined by the observer; it may be a new mRNA transcript, a specific protein, total cellular mRNA, or the nucleotides or aminoacyl tRNAs used to support the biosynthesis of these molecules. The compartment may be an identifiable, physical unit, such as the blood, the brain, or the liver, or it may be a kinetic term needed to describe the behavior of the system. For instance, if two kinds of mRNA coexist in the same cell as distinct units, then each one could be treated as comprising a compartment or pool. However, compartmental analysis was first applied to analysis of the accumulation of drugs and radioactive tracers in the body, and it was found that accumulation and decay curves do not strictly correspond to movement among physical compartments. Equally important was the rate of metabolism of the traced material by specific processes. Curves that describe the accumulation and decay of nutrients and drugs in the blood are often treated as a sum of exponential terms that include rates of metabolism, excretion, and movement into physical compartments.

Compartmental analysis functions best in quantitative studies of molecules within a cell *population*. Treating protein synthesis as though mRNA or protein in a single cell were under study has little biological relevance, because the typical organ contains approximately 10^8 cells per gram, and samples required for most biochemical analysis represent thousands of cells. The behavior of a single cell is irrelevant to most mammalian traits; what is of interest is how the combined gene products in a cell population contribute to the economy of the organism. In the absence of entraining agents such as hormones, mitotic signals, or altered membrane potential, any oscillations within individual cells would be masked. In this view, a phenotype is a statistical trait, and the investigator recognizes that an experiment is an exercise in estimating a population mean through statistical methods. Compartmental analysis seeks to estimate parameters that describe the dynamics of identifiable populations or "pools" that show uniform behavior.

There are two basic kinds of compartmental modeling. One method begins with a conceptual model such as the diagram shown in Fig. 1.1, and derives equations to fit the model structure. This is called mechanistic or physiological modeling, because it makes predictions based on the organization of the tissues, physical compartments, and transfer rates that are observed in the human or animal subject. The second method begins with a set of experimental data, such as accumulation and disappearance of an isotopic tracer in the blood plasma, and asks what set of equations and parameters would best fit the data. In that case, the commonest procedure is to fit the accumulation curve to a set of exponential functions. In this method, there is no essential correlation between the individual functions and an organ, tissue space, or fluid. This is a kinetic pool, or "a quantity of a substance which has a uniform and distinguishable kinetics of transformation or transport."

KINETIC BEHAVIOR OF A SINGLE COMPARTMENT

The single-compartmental, quantitative model of enzyme induction can be viewed as a kinetic unit of function. It refers to a product that is formed at a fixed rate under initial conditions, and that is eliminated at a rate that is proportional to concentration; these are defined as zero-order filling and first-order decay, and the process may be depicted as follows:

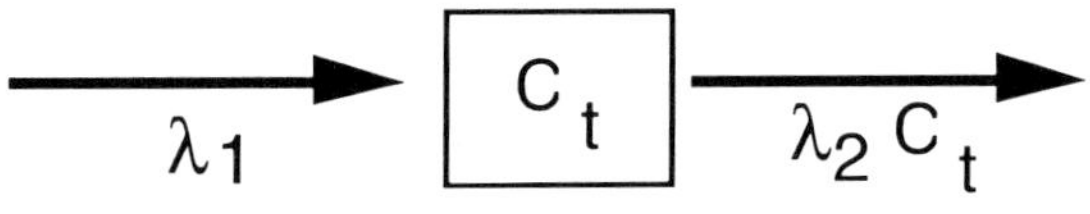

In this model, a molecule has a concentration at any moment designated C_t, and is being produced at a rate of synthesis defined as λ_1, with units of molecules produced per unit of time and tissue mass. The rate of decay, however, is proportional to concentration and is calculated by multiplying the concentration times the first-order rate constant for decay, λ_2, which has units of $time^{-1}$.

This quantitative model was introduced in the context of enzyme induction by Monod, Pappenheimer, and Cohen-Bazire[3] during studies of the effects of lactose analogs on β-galactosidase in *Escherichia coli.* They assumed that synthesis of this enzyme initially occurred at a constant rate (zero order kinetics) that could be increased in the presence of substrate when glucose was limiting for growth, and that enzyme decay was proportional to concentration (first order). When the assumptions of zero order synthesis and first order decay are valid, product will accumulate according to a simple, exponential function that can be expressed as follows:

$$C=\frac{\lambda_1}{\lambda_2}(1-e^{-\lambda}2') \qquad \text{Eq.1.1}$$

This equation predicts that, at equilibrium, the concentration (C) of b-galactosidase equals the ratio between its rate of synthesis (l_1) and the rate constant for decay (l_2). In this example, the rate of synthesis had units of enzyme activity produced per milligram of bacterial protein per unit of time. The rate constant for decay equals the fraction of the existing enzyme that is removed during each interval of time, and has units of $time^{-1}$; dividing the rate of enzyme synthesis by the fractional rate of degradation gives units of enzyme per mg of protein. If one of the kinetic parameters changes, the concentration will approach a new steady-state in an exponential fashion, with a period defined by the rate constant for decay. The exponential term, $1\text{-}e^{-\lambda_2 t}$, is a fraction that varies between the values of zero and one.

The quantitative model used by Monod and colleagues[3] is an example of a single building block of the technique called compartmen-

p. 5, par. 3 (lines 2 and 3 below equation)—
"b" in "b-galactosidase" should be Greek β;
similarly, l_1 and l_2 should be Greek λ_1 and λ_2.

tal analysis. In this case, the single "compartment" was a cell culture, and the material in the compartment was all the enzymatically active β-galactosidase. The form of systems theory that employs compartmental analysis simply establishes the flow of material into and out of each distinct, kinetic compartment of interest to the experimentalist, and produces a set of equations that describes the entire set of interactions. The technique is applicable in molecular biology because the same kinetic model was later used to analyze changes in concentrations of enzymes and mRNAs in mammals, birds, and insects.[4-6] When the relationship between mRNA production and protein accumulation is examined,[5] a two-compartment model is produced. As shown in Figure 1.1, this idea may be readily extended to include nuclear and cytoplasmic processes, as well as the external stimuli that alter gene expression in the system.

The idea that decay constants figure importantly in determining the period required for enzyme induction was recognized by Berlin and Schimke[4] during comparative studies of enzymes in mammalian liver. They concluded that the period required for several enzymes to shift to new levels after hormonal stimulation was related to their half-lives, even when rates of synthesis were similar. According to this kinetic model, *the decay constant is equivalent to the time constant for change in product concentration*, because the period required to achieve half of the total shift in concentration between two steady-state levels cannot be less than the half-life of the mRNA or protein under study. This idea may be restated in terms of compartmental analysis very simply by saying that the chemical half-life is one measure of the fractional transfer rate from the compartment of active mRNA or protein to the "compartment of oblivion." Because the half-lives of these molecules vary from a few minutes to several days, the period required to achieve a new steady state varies within the same range.

The models for interpreting mRNA and protein induction as a function of time are not difficult to unify in principle. In the simplest case, one simply assumes that the rate of protein synthesis is proportional to mRNA concentration, and seeks to quantify the number of protein molecules synthesized per mRNA per unit of time. When this is done, one has begun with what is technically called a single-compartmental model, and created a dual-compartmental, mathematical model of gene expression. That gene expression could be modeled by the technique of *compartmental analysis* was recognized by David Hoel.[5] Theorists recognize this as one of the commonest general techniques of systems theory, and basically the same model was used by other mathematicians[10-12] before and after Hoel's analysis.

The present work will follow this precedent for two reasons. Firstly, compartmental analysis is extremely straightforward and has been used by nearly all experimental scientists who have attempted to analyze gene expression quantitatively. Secondly, some of the simplest, but most powerful, microcomputer programs available for mathematical modeling are based on this method, and can be used appropriately by most biologists. Readers who learn this method are advised to always justify the underlying assumptions on which their models are based, and to locate collaborators who are trained in advanced methods of numerical analysis if unusual behavior is encountered.

ASSUMPTIONS AND NOMENCLATURE USED IN MODELING SYSTEMS

One important reason for modeling is to formulate and test hypotheses by determining whether observed experimental results fit with predictions based on current understanding of how a biological system operates. The modeler typically compares the highest quality experimental data for the system of interest to the response curves predicted from the best available quantitative theory. Inevitably, discrepancies are observed that provide evidence for unsuspected interactions, ways to improve the theory, or both, and the modeler seeks to improve his understanding of the system by improving the experimental techniques and the mathematical theory. Many current computer programs also provide statistical techniques for estimating parameters that provide the best fit between a conceptual model and the measured data.

Systems theory includes models that may either be deterministic or stochastic. *Deterministic behavior* implies that the future state of a system is predictable based on knowledge of its current state and any future inputs, whereas *stochastic behavior* includes an element of randomness, heterogeneity, or uncertainty about the future state of the system. The idea that a tenfold increase in mRNA concentration will produce a tenfold increase in the concentration of the corresponding protein is an example of a deterministic model that is also *linear* because a change in the concentration of mRNA produces an equivalent change in protein when the system attains an equilibrium. Behavior is linear if a change in input produces a proportional change in output. Although the assumption of linear behavior is seldom questioned, most biological processes are non-linear if examined over an appropriate dynamic range. For example, enzymes and transport systems become saturated as substrate concentration increases, and this produces a *non-linear response curve* typified by hyperbolic, Michaelis-Menten kinetics. If a tenfold increase in mRNA concentration only produced a six-fold increase in protein, then it is possible that the system is behaving non-linearly, or that it is not in equilibrium. The reason that compartmental analysis works very well for analyzing genetic systems is partly attributable to the fact that an increase in transcription often does produce a proportional increase in mRNA concentration, and that this is followed by a proportional increase in protein synthesis. If one considers the complexity of the process, this is surprising, but cell-free translation of mRNA has shown that responses are linear across a moderate range. In one of the original systems for cell-free protein synthesis, Pelham and Jackson[13] showed that globin mRNA was translated in proportion to concentration at ranges up to 10 μg/ml, and Palmiter[14] obtained a linear response for ovalbumin synthesis at mRNA concentrations as high as 200 μg/ml. The response to mRNA in vivo is more difficult to ascertain, but Richter and Smith[15] observed a linear increase in globin synthesis in oocytes from *Xenopus laevis* when quantities of 5-100 ng of mRNA were injected per egg.

Most biologists assume that a system of interest can be manipulated to produce an initial *steady state*, which is defined as a condition in which the material of interest is produced at the same rate it is degraded or eliminated. When this is true, the concentration of material will be constant, or time-invariant. For example, cells may be grown in a defined tissue culture medium until they reach confluence, so that a particular mRNA is maintained at a fixed concentration. The initial conditions are then changed by the addition of a hormone, cytokine, or nutrient, and the mRNA concentration is measured as a function of time. Although true steady states do not occur in biological systems, the assumption facilitates model-based analysis and is implicit in most of the literature in molecular biology. After the system has been perturbed, the mRNA (or other dependent variable) is assumed to approach a new steady state that depends on the concentration of the inducing stimulus.

TRANSFER RATES BETWEEN COMPARTMENTS ARE RELATED TO HALF-LIVES

Visualize a system of gene expression in which an investigator had treated cells with a specific inducing agent, and was attempting to relate the accumulation of mRNA in the cytoplasm to the rates of transcription; nuclear processing, decay, and transport; and cytoplasmic degradation. It should be self-evident that the rate of transcription responds first, and could come to a new equilibrium before the nuclear transcript attained its new steady-state level; likewise, the concentration of mature mRNA in the cytoplasm could only reach an equilibrium after the preceding processes had attained new, constant rates. This "thought experiment" exemplifies the most basic tenet of compartmental analysis: *rates of transfer between compartments determine the time required to approach a new steady state.* Note that it is not necessary to actually achieve the steady state; the model suggests that transfer rates determine the rate of approach to steady state whether or not there is time to attain the new value.

Observe also that (1) rates of degradation or metabolism are equivalent to rates of transfer, transport, processing, or secretion in compartmental analysis, and that (2) these rates combine to determine the time course of response for the system. In compartmental analysis, individual compartments are diagrammed as circles or rectangles with arrows connecting each one to the next compartment or to the rest of the system. Each transfer function has the units of reciprocal time, or amount of material transferred per unit of time. In other words, each arrow represents a rate coefficient, or fractional transfer rate. *Fractional transfer rates are not kinetic constants*; they simply indicate how much of the material in the pool moves out of the pool by the indicated route during any interval. Usually, these rates are treated as constants, but this is not essential to compartmental analysis. If the investigator determines that the transfer rate varies as a function of concentration or time after a stimulus, then it is very easy to modify the program to indicate the correct relationship with a hyperbolic curve or sliding function.

As the fraction of material transferred or eliminated from a compartment per time interval increases, the shorter the half-life becomes, and the quicker the system approaches its new equilibrium. Therefore, the fractional transfer rate (or fractional elimination rate, λ) is inversely proportional to the half-life ($t_{1/2}$). By definition, the relationship is stated as follows:

$$\lambda = 0.693/t_{1/2} \qquad \text{Eq.1.2}$$

The meaning of this equation is that material that is transferred or degraded rapidly has a short half-life. In other words, half-life is a derived property that is related to the stability or persistence of a mRNA or protein in the cell, presumably because each protein and mRNA interact with cellular systems for degradation or transport to differing degrees.

COMPARTMENTAL MODELS ARE SELF-ASSEMBLING AND PREDICT STEADY STATE RELATIONSHIPS

An important attribute of compartmental modeling is that it permits the investigator to build a system using a flow chart or set of block diagrams. After rates have been assigned, the steady state relationships may be written down with little practice, and can be solved using a hand-held calculator. No knowledge of differential equations is necessary to do this, because in the initial or final steady state, there is no change. This is a very useful method for individuals who think better within a visual context than in a numerical context. The diagram is constructed on the computer screen using the icon-based tools provided by the simulation software, and numerical values or equations are typed in at the keyboard by way of dialog boxes. This method overcomes fear of mathematics, because the investigator is working visually, and the computer itself is fearless. All the investigator needs to provide is a diagram linking the components under investigation, the rates of transfer, and a notion about whether the components are related in a linear or non-linear manner.

THE RELATIONSHIP BETWEEN SIMULATION AND EXPERIMENTATION

Systems theory may be defined as a formal means of asking how identifiable components in a complex system are quantitatively and temporally related to specific traits or responses to change. The importance of systems theory as a tool is not to exactly mimic nature or a particular experimental outcome, although it is gratifying when this happens; instead, it is to provide a quantitative way to interpret results and think about experimental design. Most scientists who use systems theory are interested in determining whether a postulated mechanism for control is supported by experimental data, which requires a method to generate quantitative predictions (simulation), and a method to acquire appropriate experimental data. In its simplest sense, this involves asking how to calculate initial quantities and rates, and the shape of curves that describe changes over time. The curves are familiar: they may be hyperbolic, simple exponentials, sigmoidal, or perhaps represent the sum of several such curves. Typically, values may oscillate, as in the case of circadian rhythms in the cell cycle, hormone secretion, or gene expression. Systems theory is a way of generating similar curves using tools that include differential equations, linear and non-linear algebra, and other forms of numerical analysis. When this set of techniques is applied to a specific system, the result is a quantitative model that can be used as a means of comparing experimental results with predicted outcomes obtained by computer simulation.

Because few biologists have a working knowledge of kinetic modeling, use of systems theory traditionally required collaboration between experimentalists and mathematicians. New programs for personal computers circumvent this relationship to some extent by providing an interface that lets the experimentalist focus on ideas, while providing solutions to the systems diagram through solidly based numerical methods. Computer programs that are designed for simulation provide the user with at least three components: (1) an interface that permits any user to set up quantitative relationships that predict a system's future behavior, (2) one or more methods (numerical routines or algorithms) for solving the equations, and (3) methods to start and stop the simulation, to view the results, and to save the data for future use. Research-oriented simulation programs also include statistical subroutines that enable the investigator to compare simulation results with experimental data, and that enable parameters that give the best fit to be estimated when the form of an equation is known. Several computer programs that are available through federal agencies and private companies are described at the end of this chapter.

RESOURCES FOR COMPUTER SIMULATION

Numerous programs for computer simulation have been developed by workers in fields as diverse as economics, engineering, pharmacodynamics, public health, and ecology.[16-19] The different simulation programs are akin to the variety of word processors; each one has a range of capabilities and is evolving, and it is likely that several will be actively maintained for the indefinite future. Just as word processors removed the fear of making typographical errors, programs for simulation remove the fear of making mathematical errors. If a routine generated in a simulation program gives useful data, it can be saved and reused in the future; if it does not produce useful results, it can be modified or discarded. Mistakes cost nothing, as long as the experimentalist is careful about not overinterpreting the results, and troubleshooting can function as a useful exercise in thinking.

In preparing this volume, the program called Systems Thinking, Experiential Learning Laboratory with Animation (STELLA II) was chosen because of its ease of operation and because an inexpensive version is available that is offered on an accompanying diskette. This program evolved from work initiated by Jay Forrester and colleagues for analysis of issues in worldwide macroeconomics and resource allocation.[17-19] Table 1.1 lists a few other examples from the many possible selections.

A variety of general simulation programs has been developed under the sponsorship of the Biomedical Research Technology Program,[20] which funds several resource centers and publishes a catalog of products useful for biomedical computing. One of the first general modeling programs is called Simulation, Analysis and Modeling (SAAM), which was developed at the National Institutes of Health by Mones Berman.[21] The original versions were difficult to learn, but a modified version called Conversational SAAM (CONSAM) is more user-friendly and has been widely used by biomedical scientists, especially for studies involving isotopic tracers. A new program called SAAM II has been developed that promises to include the capabilities of SAAM and the ease of use of a menu-driven, icon-based interface. A program called Simulation Control Program (SCoP) was developed to facilitate physiological and biochemical modeling.[22] It has been used by Kohn and colleagues to produce a model that examines the interactions of dioxin and estrogen on cytochrome P450 enzymes in relation to risk for hepatic cancers.[23,24]

Modeling differs from the typical process of concept development in that a hypothesis may be stated in quantitative or mathematical terms using a computer program as an aid to developing a diagram or statement of relationships. Several new programs, such as STELLA II, CONSAM, SAAM II and SCoP are organized to facilitate setting up

Table 1.1. Examples of software for computer-assisted modeling

Software package	Source	User requirements	Computer requirements	Telephone/ e-mail
SAAM II	RFKA, Seattle, WA	Icon-based language	Sun, IBM, Macintosh	800-421-SAAM
CONSAM	Lab Math. Biol., NIH, Bethesda, MD	Learn SAAM language	Sun, DEC WS, IBM 486	301-496-8914
LYSIS ADAPT II	Biomed. Sim. Res., Los Angeles, CA	Learn to interact with packages	VAX, SUN, DOS	213-740-0839
SCoP, SCoPfit	Sim. Resources, Berrien Springs, MI	Write and enter equations; use library	IBM, UNIX	Kootsey@peter. cs.andrews edu
STELLA II	High Performance Systems, Hanover, NH	Use icon-based interface, finite difference eqs.	Macintosh or IBM with Windows	800-332-1202

block diagrams on the computer workstation, with all parts of the diagram interconnected by quantitative relationships. The latter three programs also provide built-in curve fitting routines for parameter estimation. The diagrams can be completed in a matter of minutes, and after the quantitative relationships have been supplied, the result is a computer program that immediately permits simulations to be performed. If the experimentalist has an accurate concept of the system, then the simulation should closely match the experimental results, and enable predictions to be made about the likely outcome of a specific change or perturbation. More commonly, discrepancies between the experimental data and the simulation suggest improved hypotheses, new experimental designs, or missing information. The great advantage of computer-assisted simulation is that it costs little and can be accomplished rapidly on a modern microcomputer. Using a computer in this way provides a new tool for thinking and encourages an understanding of relationships that are not always obvious.

By incorporating quantitative, temporal relationships into a model, it may be possible to make inferences that yield fruitful new lines of research, or to devise new ways to test ideas at the bench. If an investigator does not take advantage of this quantitative approach, then an important element in research is missing. If either the experimental data or the conceptual model is qualitative, then it is difficult to determine whether discrepancies exist between the results and the theory. The microcomputer solves this problem by enabling the investigator to employ quantitative methods so that results may be compared statistically. The following pages will describe essential concepts of systems theory in relationship to gene expression, will provide a guide to the methods and assumptions that have proved most useful to experimentalists, and describe some of the resources that are available for microcomputers.

References

1. Gleick J. Chaos, Making a New Science. New York: Penguin Books, 1987.
2. Mendel G. Versuche uber Pflanzenhybriden In: Stern C, Sherwood ER, eds. The Origin of Genetics. A Mendel Source Book, San Francisco: WH Freeman and Co, 1965.
3. Monod J, Pappenheimer AM, Cohen-Bazire G. La cinetique de la biosynthese de la β-galactosidase chez *E. coli* consideree comme fonction de la croissance. 1952; 9, 648-660.
4. Berlin CM, Schimke RT. Influence of turnover rates on responses of enzymes to cortisone. Mol. Pharmacol. 1965; 1: 149-156.
5. Hoel DG. A simple two-compartmental model applicable to enzyme regulation. J. Biol. Chem. 1970; 245:5811-5812.
6. Hargrove JL. Microcomputer assisted kinetic modeling of mammalian gene expression. FASEB J. 1993; 7:1163-1170.
7. Godfrey K, ed. In: Compartmental models and their applications. London: Academic Press, 1983.
8. Jacquez JA. Compartmental analysis in biology and medicine. 2nd ed. Ann Arbor: University of Michigan Press, 1985.

9. Green MH, Green JB. The application of compartmental analysis to research in nutrition Annu Rev Nutr 1990: 10:41-61.
10. Goodwin BC. Temporal Organization in Cells, London: Academic Press, 1963.
11. Maynard Smith J. In: Mathematical ideas in biology. Oxford, Cambridge University Press, 1968.
12. Yagil G. Enzyme induction. In: Segel LA, ed. Mathematical models in molecular and cellular biology. Cambridge: Cambridge University Press, 1980: 68-83.
13. Pelham, HRB, Jackson RJ. An efficient mRNA-dependent translation system form reticulocyte lysates. Eur J Biochem 1976; 67:247-256.
14. Palmiter RD. Ovalbumin mRNA translation. J Biol Chem 1973; 248:2095-2106.
15. Richter JD, Smith LD. Differential capacity for translation and lack of competition between mRNAs that segregate to free and membrane-bound polysomes. Cell 1981; 27:183-191.
16. Collins JC. Resources for getting started in modeling. J Nutr 1992; 122:695-700.
17. Forrester JW. Urban Dynamics Cambridge MA: MIT Press, 1969.
18. Meadows DH, Meadows DL, Randers J et al, eds. In: The limits to growth. New York: Universe Books, 1972.
19. Hayes B. Balanced on a pencil point. Amer. Scientist 1993; 81:510-516.
20. Department of Health and Human Services. Biomedical Research Technology Resources: A Research Resources Directory, 8th ed, Rockville MD: Research Resources Information Center, 1990.
21. Berman M, Grundy SM, Howard BV, eds: Lipoprotein kinetics and modeling, New York: Academic Press, 1982.
22. Kootsey JM, Kohn MC, Feezor MD, Mitchell GR, Fletcher PR. SCoP: An interactive simulation control program for micro- and minicomputers. Bull Math Biol 1986; 48:427-441.
23. Kohn MC, Portier CJ. Effects of the mechanism of receptor mediated gene expression on the shape of the dose-response curve. Risk Analysis 1993; 13:565-572.
24. Kohn MC, Lucier GW, Clark GC, Sewall C, Tritscher AM, Portier CJ. A mechanistic model of effects of dioxin on gene expression in the rat liver. Toxicol. Appl Pharmacol 1993; 120, 138-154.

CHAPTER 2

Time and Quantity in Gene Expression

If you sit with a beautiful girl, two hours seem like two minutes. If you sit on a hot stove, two minutes seem like two hours. That's relativity.
—Albert Einstein

The organism itself frequently becomes the coordinate system on which time scales are based.
—Harold Boxenbaum

THE TIMING OF GENE EXPRESSION

Time is too often taken for granted. Although texts in biochemistry and genetics discuss reaction *rates*, the topic of time is not indexed and is seldom discussed, as though it were a physical substance that is identical for all organisms. However, for enzyme adaptation to be effective, new gene products must be synthesized in appropriate concentrations during periods that are physiologically relevant. Given that a mammalian cell is about 1,000 times larger than a bacterial cell and has a 20-fold longer cell division time,[1] it is clear that the problem of time scale is not the same for these organisms; in fact, it is not likely to be the same for different mammals. This is biological relativity, a concept recently discussed by Boxenbaum[2] in terms of allometric scaling. The mouse was chosen for genetic studies due to its short life span and rapid sexual maturation, and not incidentally, its heart beats over 500 times per minute, in comparison to about 72 beats per minute for humans. What does this have to do with the genes? Simply put, it means that nutrients and drugs are delivered to the tissues more rapidly in mice than in men, and it has been demonstrated that both heart rate and the rate of drug disposition can be scaled to body weight for numerous species.[2] Drug metabolism is controlled by a variety of inducible enzymes. If the rate of drug elimination is related to heart rate and body weight, is it possible that the rate of enzyme induction is also scaled in this way?

Aging is another example of a process whose rate differs for each species and is influenced by the genes, as are individual processes such as the cell cycle, growth, development, and loss of vitality. In fact, it is arguable that the most basic mathematical function in biology is the Gompertz-Makeham mortality curve, which is commonly described by two simple terms; one that is age dependent, and one that is not:[3]

$$R_m = A + R_0 e^{-\alpha t} \qquad \text{Eq. 2.1}$$

In this equation, the mortality rate at any age, R_m, is the sum of an age-independent constant, A, plus the initial number of individuals in the population, R_0, multiplied times an exponential function. Aging is a phenotype, and systems theory, or mathematical modeling, provides a means to begin thinking about time-dependent problems quantitatively. To leap to a conclusion, if it is true that the rate of degradation of mRNAs and proteins is also allometrically scaled, then it could be argued that these processes play a pivotal role in aging and possibly in the sense of time.

THE FLOW OF GENETIC INFORMATION

The term, flow, implies rate, and the use of systems theory to study gene expression requires that the problem be phrased in quantitative terms. In 1970, Francis Crick[4] restated two fundamental, qualitative ideas concerning the transfer of biological information. The first idea was called the Sequence Hypothesis, which "in its simplest form assumes that the specificity of a piece of nucleic acid is expressed solely by the sequence of its bases, and that this sequence is a code for the amino acid sequence of a particular protein." The second concept was called the Central Dogma of molecular biology, which "states that once 'information' has passed into protein it cannot get out again. Information here means the precise determination of sequence, either of bases in the nucleic acid or of amino acid residues in the protein."

In this statement of the laws that govern transfer and expression of genetic information, several points were omitted that can be addressed using systems theory. First, the matter of *scale*, or relative quantities of DNA, RNA, and protein in cells was not discussed, although it was known that information was amplified because the machinery required to effect information transfer possessed a variety of enzymatic activities. Second, it was evident to Crick, as an accomplished protein crystallographer, that the amino acid sequence of a protein could represent a set of instructions for folding the peptide chain in order to achieve a specific capacity for ligand binding or enzymatic activity. If the protein failed to fold properly, the transfer of genetic information would fail even though the linear, encoded sequence had been properly translated. Third, a protein was not identical to a phenotype, but could be viewed as a means to generate a phenotype if the protein acquired its capacity and acted in concert with other cellular constituents. Systems theory requires that gene expression should be viewed in its entirety, and that quantities and rates be determined. Historically, the major advances in understanding the processes of gene expression took place when new techniques that permitted precise measurement of these quantities were developed, and not incidentally provided means to obtain the complete code for each kind of linear information and insights into higher order structure. In order to illustrate how a quantitative approach can augment thinking about genetic systems, the chapter will discuss the effects of inducing agents on the ovalbumin gene, the *lac* operon in *Escherichia coli* and the mammalian enzyme, tryptophan dioxygenase (E.C. 1.13.11.11).

QUANTIFYING GENE EXPRESSION

The steroidal regulation of the gene for ovalbumin provides a classic example of the quantitative approach to gene expression. As summarized by Palmiter,[5] several groups devised immunological techniques for purifying and quantifying ovalbumin, for isolating polyribosomes engaged in ovalbumin synthesis, and for estimating ovalbumin mRNA quantity by translation and by hybridization to complementary DNA. In a culminating paper, McKnight and Palmiter devised a method to measure transcription of ovalbumin mRNA in isolated nuclei.[6] Together, these techniques enabled Palmiter to estimate the rates of all the processes involved in expression of the estrogen-dependent ovalbumin gene in the uninduced and fully induced conditions; this is an excellent example of the power of steady-state analysis.

The analysis showed that each single-copy gene was transcribed at a rate of 11-17 transcripts per minute in estrogen-stimulated cells, giving a rate of synthesis of 22-34 new mRNAs per minute for the two genes present in diploid cells. The mRNA was degraded with a half-life of 24 hours, and accumulated to a level as high as 101,000 molecules per cell. The average mRNA contained 12 ribosomes, which initiated at a rate of 9.2 ribosomes per minute, and generated polypeptide chains at a rate of 5 amino acids per second, to yield a transit time for completion of the ovalbumin polypeptide of 1.3 minute. These data showed that the rate of ovalbumin synthesis reached peak values of 600,000 molecules per minute per cell after 5-6 days of estrogen stimulation, and this rate was accounted for by the transcription of two genes per cell at about 35% of maximal efficiency. This classic analysis of vertebrate gene expression demonstrated the power of quantitative techniques to explore the way that genetic activation could generate a cell-specific phenotype. The information available is sufficient to enable kinetic modeling of the process, and to determine whether the predicted timing also agrees with the observed, developmental acquisition of the ability to secrete ovalbumin. The steady-state analysis showed that the mature chicken oviduct indeed produced the amount of ovalbumin found in chicken eggs, and along the way, gave some of the first evidence of the existence of intervening sequences and a widely used technique for nuclear run-on transcription.

THE TIMING OF GENE EXPRESSION IN BACTERIA

The production of a protein with an enzymatic activity may be the simplest example of a phenotype. In order to illustrate the temporal aspect of gene expression, or time required for a specific phenotype to become fully developed, it is useful to compare the first inducible genetic systems to be discovered and characterized in bacteria and mammals. The structural elements involved in regulating these two genes are shown in Figures. 2.1. The concept that genes were regulated by elements located nearby on the same chromosome ("operators") was developed in studies on sites that control lactose utilization (*lac* operon) in *E. coli* by Jacob, Monod, and colleagues.[7-11] The *lac* operon is controlled by two regulatory proteins, or *trans*-acting factors, which bind to specific DNA sequences near the binding site for RNA polymerase, or promoter (Fig. 2.1). When lactose is not present, the structural gene for the lac repressor (*lacI*) is transcribed, and the repressor binds to the operator as a tetramer of four identical subunits; this blocks binding of RNA polymerase to the promoter for the *lac*

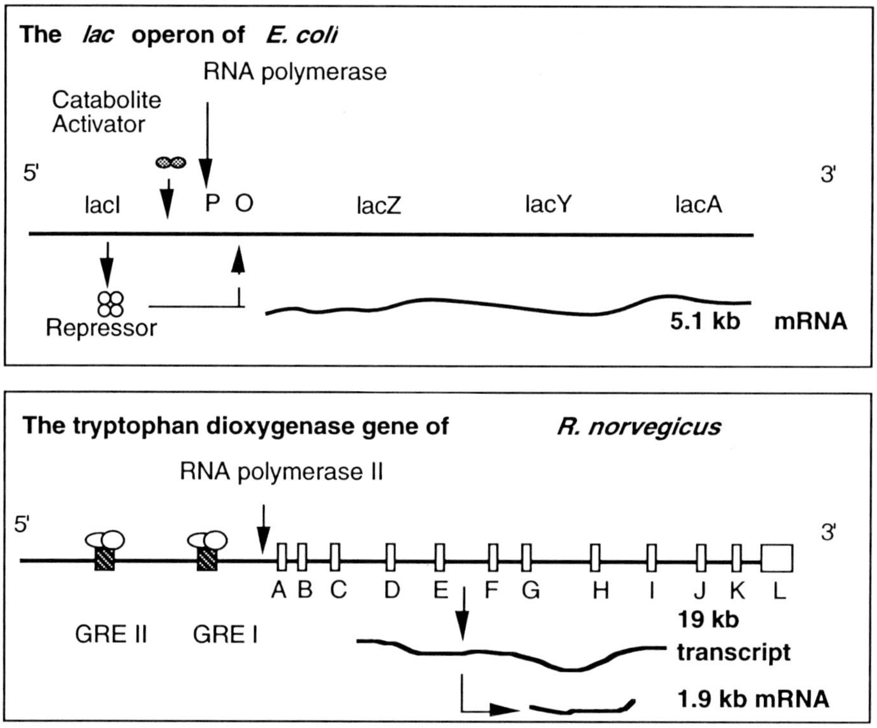

Fig. 2.1. Top: The lac operon of E. coli includes the operator (O), a regulatory DNA segment that binds the lac repressor and prevents transcription of a 5.1 kb mRNA which encodes three open reading frames for the proteins, β-galactosidase, lactose permease, and transacetylase. Transcription is initiated in the presence of the inducer, allolactose, which binds to the lac repressor and causes it to be released from the operator. Maximal rates of transcription also require that cyclic AMP be present, which binds to the catabolite activator protein and enhances the binding of RNA polymerase to the promoter (P). Bottom: The mammalian gene for tryptophan dioxygenase encodes a 19 kb primary mRNA transcript that includes 12 exons (A to J) and 11 introns. Transcription is regulated by tissue-specific mechanisms and by at least two glucocorticoid response elements (GRE I and II). The rate of transcription increases by 4-10 fold when glucocorticoid hormones activate the glucocorticoid receptor and saturate enhancer elements at GRE I and II. The 19 kb primary transcript is then processed to generate a mature mRNA of about 1.9 kb with poly[A] tail included.

Table 2.1. Typical values for parameters and rates needed to quantify gene expression in E. coli and mammalian cells

	E. coli	**Mammals**
Initiation	1 sec^{-1}	10 min^{-1}
Transcription	2500 nt min^{-1}	3,000 nt min^{-1}
RNA processing	not applicable	ca. 10 min
Half-life, nuclear RNA	not applicable	ca. 10 min
Nucleocytoplasmic transport	not applicable	ca. 10 min
Half-life, mRNA	1-3 min	1-20 h
mRNA translation	2700 nt min^{-1}	720 nt min^{-1}
Half-life, protein	20-60 min	2-100 h

operon. The second protein required for control of the *lac* operon is the catabolite activator protein, which is not capable of binding with high affinity when glucose is present.

When lactose is added to cultures of *E. coli* in the absence of glucose, the sugar permeates into the cell and is converted enzymatically to an inducing agent, allolactose (1,6-O-β-D-galactopyranosyl-D-glucose). The action of the inducer can be mimicked artificially by adding an analog, isopropylthiogalactoside (IPTG). This substance permeates into the cell and acts within a few seconds to remove the *lac* repressor and promote initiation of transcription. However, the maximal rate of transcription is only attained when glucose concentration is low and the concentration of cyclic adenosine 3',5'-monophosphate (cyclic AMP) is elevated. Cyclic AMP binds to the catabolite activator protein, and enables it to occupy the promoter adjacent to the RNA polymerase binding site (Fig. 2.1). Presence of catabolite activator protein greatly facilitates the transcription of this operon when the lac repressor has been removed.

The *lac* transcription unit spans 5,100 nucleotides and encodes a polycistronic mRNA that includes three genes, β-galactosidase, lactose permease, and transacetylase; the *lacZYA* transcription unit encodes three enzymes designated *lacZ*, *lacY*, and *lacA*, respectively. The majority of the delay that occurs after inducer is added and before mRNA is detectable results from the time needed to complete transcription, about 2 minutes at a rate of about 2,500 nt per minute (Table 2.1). The messenger RNA for β-galactosidase increases from an undetectable level to a new plateau within approximately 4-5 minutes.[8-10] In response to the increase in mRNA, synthesis of β-galactosidase rapidly increases. Because protein synthesis is coupled to transcription, ribosomes initiate protein synthesis while the nascent mRNA is still being transcribed. The rate of translation is about the same as the rate of transcription. Because the *lacZ* gene is 3510 nt in length, the first monomers of β-galactosidase are produced within 1-2 minutes after translation initiates, and about 3-5 minutes after transcription initiates (Fig. 2.2). In exponentially growing cultures of bacteria, the enzyme accumulates to a very high level, with half-maximal induction taking place between 1-2 hours.[10] The maximum quantity of β-galactosidase increases 1000-fold, resulting in about 60,000 monomers per cell (or 15,000 catalytically active tetramers). These data suggest that *E. coli* can acquire the phenotype of lactose utilization within about 10 minutes of exposure to a nutritional substrate, but that the maximal rate of utilization requires 30-40 minutes to develop.[11] Even though the maximum rate of β-galactosidase synthesis is attained within 5 minutes, the full expres-

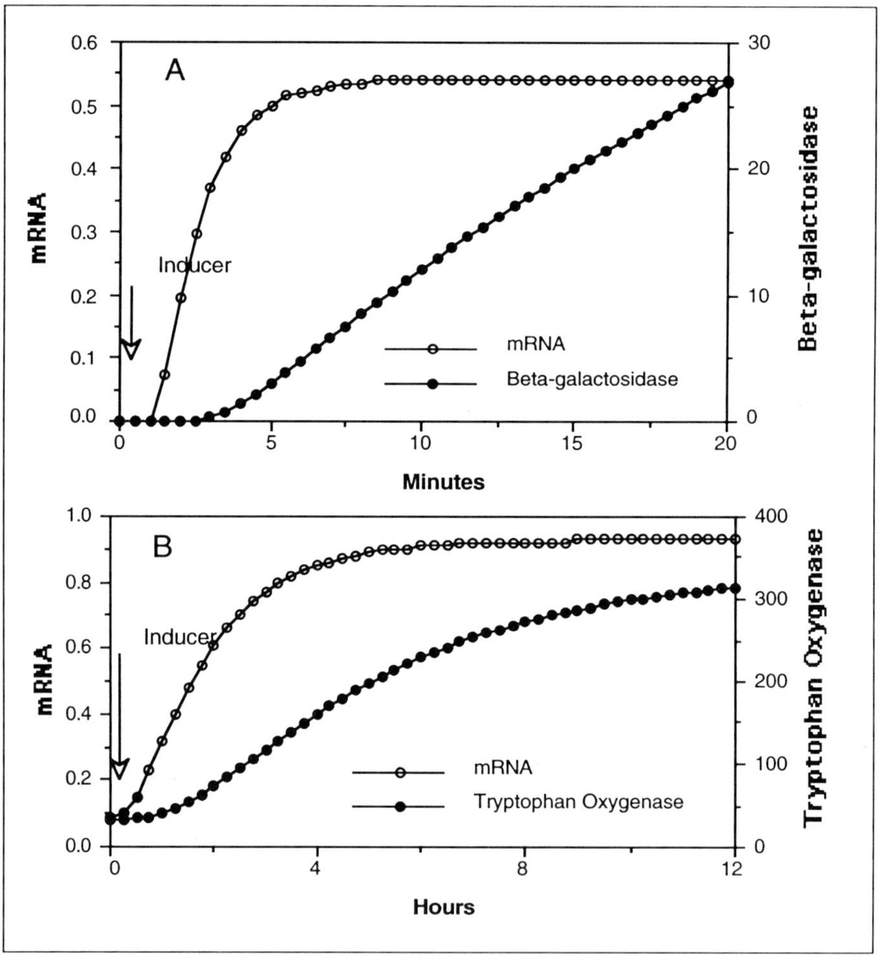

Fig. 2.2. Top: Time course of induction for the lac operon showing the increase in mRNA within 2 minutes when an inducing agent (isopropylthiogalactoside) is added (arrow). The mRNA concentration attains approaches a new steady state within 5 minutes and causes a proportional increase in synthesis of β-galactosidase that begins within 3-5 minutes. Bottom: Transcription of the tryptophan dioxygenase gene increases within minutes when rats are injected with glucocorticoid hormones (arrow), and the mature mRNA begins to accumulate in the cytoplasm within 30 minutes. The rate of protein synthesis attains a maximum value after about 3 hours, and tryptophan dioxygenase activity attains peak values 5-8 hours after the glucocorticoid is administered.

sion of the phenotype necessitates that the enzyme accumulate to its maximal level and also that lactose uptake be increased by expression of the permease enzyme.

THE TIMING OF ADAPTIVE GENE EXPRESSION IN MAMMALS

The first clear evidence that mammalian metabolism could be regulated at the genetic level by nutritional substrates and hormones was provided by the discovery that tryptophan and adrenal glucocorticoid hormones regulated the activity of tryptophan dioxygenase in liver.[12] This enzyme initiates the utilization of tryptophan when the amino acid is present in excess over requirements for protein synthesis. This pathway is necessary for tryptophan utilization in gluconeogenesis and for endogenous synthesis of nicotinamide-containing cofactors, and deficiency in the dioxygenase would presumably be correlated with enhanced susceptibility to development of pellagra. Enhanced activity of the dioxygenase causes less tryptophan to be available for synthesis of serotonin or the neural modulators, kynurenic acid and quinolinic acid. Diminished serotonin due to enhanced metabolism of tryptophan in liver has been suggested to play a role in the development of several affective disorders, including Tourette syndrome, and to be associated with alcoholism.[13,14]

The gene that encodes tryptophan dioxygenase is 19 kb in length and is interrupted by 11 intervening sequences (introns) that are removed during mRNA processing to generate a mature transcript of 1.9 kb.[15-17] The gene is transcriptionally regulated in part by binding of activated glucocorticoid receptor to two glucocorticoid response elements located in the 5' flanking segment (Fig. 2.1). In contrast to the rapid increase in lactose utilization by bacteria, studies with tryptophan dioxygenase demonstrated that steroids such as cortisol or dexamethasone caused the enzyme to increase from a low but detectable level to a new maximum after 8-12 hours (Fig. 2.2).[12] Adrenal steroids increase the rate of synthesis for tryptophan dioxygenase by four to tenfold, and tryptophan stabilizes the enzyme to facilitate a much larger increase in enzyme concentration. When specific assays for the messenger RNA were developed, workers were able to demonstrate that glucocorticoid hormones increased the rate of transcription for the tryptophan dioxygenase gene within 20-30 minutes after intraperitoneal injection, and that mRNA accumulated to a new plateau during 3-4 hours.[17] Whereas enzyme induction in bacteria can attain maximal levels within 30-40 minutes, the process in mammals requires several hours, and in some cases may require several days to complete. How can the remarkable difference in the time course of enzyme adaptation in bacteria and mammals be explained?

PHYSICAL BASIS FOR THE TIME COURSE OF GENE EXPRESSION

Genes are not themselves catalysts, but produce enzymes at concentrations that are effective in controlling the use of energy, organic, and inorganic molecules to maintain growth, metabolism, and other life processes. Cells have finite sizes, and each cell must maintain a thousand or more proteins; therefore, it is critical to adjust the rates of transcription, translation, and turnover so that metabolic reactions are balanced. This task differs for prokaryotes and eukaryotes because the concentration of individual genes is not the same in bacteria and mammalian cells. The typical cell volume for growing *E. coli* is about 1 μm^3,[1] whereas the volume of a typical liver cell is about 11,000 μm.[3,18] This means that products generated from haploid, bacterial genes only need to fill a space that is 1/5,000th as large as the volume regulated by diploid, mammalian genes. The rate of transcription is very similar for bacterial and mammalian genes, and subsequent nuclear events take place within a few minutes in eukaryotes (Table 2.1). Based on the size of the two genes shown in Figure 2.1, the lac operon can be transcribed in under 2 minutes, whereas the tryptophan oxygenase gene requires about 6 minutes for a round of transcription. This has little bearing on the timing of gene expression, because the rate is controlled principally by transcriptional initiation. The translational efficiency of prokaryotes is about three fold greater than that of eukaryotes. This gives bacteria a three fold net advantage in rates of transcription and translation, but the small size of the bacterial cell enables it to achieve a high concentration of catalysts rapidly, which is important because the velocities of enzyme-catalyzed reactions are proportional to enzyme concentrations.[19]

According to the compartmental, kinetic model, the chemical stability of gene products controls the time course of accumulation of gene products in the cytoplasm. In contrast to the relatively long half-lives of mammalian mRNAs, which range from about 20 minutes to 100 h,[20-22] half-lives of bacterial mRNAs are typically 1-3 minutes, with an average of about 1.3 minutes.[1,23] Most proteins synthesized on the mRNA templates are relatively stable in bacteria, and this is true for β-galactosidase. However, bacterial proteins are diluted in each round of cell division, giving them an effective half-time of 20-60 minutes. The result of these separate considerations accounts for the observed time course for *E. coli* to utilize lactose at maximal rates. The initial delay seen between addition of the inducer and detection of the first transcripts includes a few seconds needed for the inducer to permeate into the cells and dislodge the lac repressor, followed by a period of about 2 minutes during which the first primary transcripts are synthesized. The mRNA then accumulates to a new plateau that is attained when the rate of synthesis is balanced by the rate of degradation. The kinetic model suggests that this should be achieved after approximately five half-lives elapse, and the observed data are in accord with the measured half-life of about 1 minute. The first β-galactosidase enzyme becomes detectable within 2 minutes after the mRNA, as expected from the rate of mRNA translation. The enzyme is stable, but becomes diluted within each cell as a result of cell division. Therefore, the effective half-life is equivalent to the generation time of 20-60 minutes.

Typical mammalian transcription units are divided by intervening sequences, and most primary transcripts range from 2-30 kb with averages of 8-10 kb, whereas processed mRNAs average about 2 kb in length. The structural gene for tryptophan dioxygenase is 19 kb in length, and would require about 6 minutes to complete after glucocorticoid hormones have activated the receptor and bound to GRE I and II (Fig. 2.1). The time needed to process the primary transcript is relatively short, with typical values that do not exceed 10 min. Accumulation curves for nuclear mRNA are characterized by half-times of 10-20 minutes, which reflects the rates of nuclear decay (half-time, about 10 minutes) and transport from the nucleus to the cytoplasm (Table 2.1). The result of all nuclear processes is that first mature mRNAs are observed in the cytoplasm within 10-30 minutes after hormonal treatment (Fig. 2.2). Thereafter, the mRNA concentration increases in a manner that is consistent with a chemical half-life of about 1 hour, and protein synthesis increases accordingly. Because the half-life of tryptophan dioxygenase is near 2 hours, the enzyme concentration attains peak values after several hours.

This comparison suggests that the difference in rates of enzyme adaptation in bacteria and mammals are not explained by the processes of transcription mRNA processing, or initiation of protein synthesis. Instead, the concentration of bacterial enzymes can change very rapidly because the effective half-lives of their mRNAs is 1-3 minutes, and the effective half-lives of bacterial proteins is 20-60 minutes, owing to the rate of cell division. In contrast, mammalian mRNAs and proteins have half-lives that range from a few hours to many days, and the mammal cannot adapt rapidly by genetic means due to the slow rate of turnover for cellular constituents. Thus, the bacterium has given up individuality for rapid adaptability, whereas the mammal has relinquished rapid genetic adaptability for stability and long life.

THE SIMPLIFYING CONCEPT OF FRACTIONAL CHANGE

The reason that rate constants for degradation control induction periods for mRNAs may be explained by considering two, simple expressions. The steady-state relationships suggest that the decay rate (λ_2) fixes the ratio between the rate of mRNA synthesis (λ_1)and the total amount of a specific mRNA in the system at equilibrium (R_{ss}):

$$\lambda_2 = \frac{\lambda_1}{R_{ss}} \qquad \text{Eq.2.2}$$

Unless the decay constant changes, the same ratio will hold during the initial and final steady-states; if the rate of transcription increases, the absolute amount of product will increase until the processes are balanced. How much time will be required to attain a new equilibrium? To provide a general answer to this question, note that when the rate of transcription changes, the amount of change in concentration of an mRNA at any moment may be expressed as a *fraction* of the total shift needed to attain the new, steady-state level:[21]

p. 23—numerator in Eq. 2.3 should read:
$R_t - R_0$, instead of $R_{ss} - R_0$, as follows:

$$f = \frac{R_t - R_0}{R_{ss} - R_0} = 1 - e^{-\lambda_2 t} \qquad \text{(Eq. 2-3)}$$

This equation expresses induction as a fraction of the total shift needed for a change in rate of transcription to produce an equivalent change in mRNA concentration. The fractional change increases exponentially from the initial value of zero to a limit of one in proportion to the unique value of the decay constant and the amount of time that has elapsed since the stimulus occurred. This fraction will equal 0.5 when elapsed time equals the half-life (or $\ln 2/\lambda_2$) of the mRNA in question. Therefore, half of the total shift in mRNA concentration will have taken place when the elapsed time equals the half-life of the mRNA, and induction will be 97% complete after five half-lives have elapsed. Because elapsed time is the same for all gene products in the system, the unique decay constant (or half-life) is the only variable that influences time course in this simple model. Processes that occur before transcription attains a new maximum are not taken into account, and the model predicts no delays prior to onset of mRNA production.

The quantitative relationship between half-lives and the fraction of the total shift completed at various times after the rate of transcription has increased to a new level is illustrated in Figure 2.3. For molecules with average half-lives (about 20 hours for mRNA in liver[20]), a small fraction would accrue towards the new steady state during one hour. In contrast, a much larger fraction would be added for unstable molecules even though the yield is lower. The equilibrium concentration of mRNA increases linearly as half-life increases, whereas the net amount of mRNA added to the pool during the first interval reaches a limit. *For indefinitely short half-lives, the limit equals the final concentration of product.* For long half-lives, the limit equals the total amount of mRNA synthesized during the period of interest; this value is always less than the final concentration. Therefore, the *ratio* between the increment produced and final concentration converges towards values of one for short half-lives and zero for long half-lives.

To give an example, if the rate of transcription for an mRNA increased from 0.1 to 1.0 pmol/h/gram of tissue, the final concentrations for mRNAs with half-lives of 0.25 and 16 hours are expected to equal 0.36 and 23 pmol/g, respectively (Eq. 2.2). For the mRNA with the short half-life, the concentration would increase by 0.3 pmol in the first hour, so induction would be 83% complete (0.3/0.36). For the stable mRNA, the concentration would increase faster (by 0.88 pmol in the first hour), but this represents only 4% if the final value (0.88/23). The chief implication of this model is that *the time course for approach to steady state is expected to be independent of the magnitude of the change in transcription or absolute quantity of product in the system, but proportional to product half-lives.*

VARIETY IN THE TIME COURSE OF MAXIMAL ENZYME INDUCTION

To what extent does the conclusion that half-life is related to the time course of induction apply to mRNAs? Analysis of induction curves for the 24 mRNAs and proteins listed in Table 2.2,[24-67] shows that the period required to achieve 50% of the new, steady-state level for these gene products is correlated with half-life (Fig. 2.4). The values plotted in Figure 2.4A were the periods needed for each mRNA to achieve half of its steady-state value; for Figure 2.4B, the criterion was the period needed for each protein to reach half of its steady-state level, based on the principle of fractional induction explained above. Half-lives for some of the gene products studied vary according to experimental treatments; hence, the values plotted were the values in effect during the induction period. Corrected rates of transcription for the mRNAs in question ranged from about 1 part per million per kilobase (ppm/kb) to 1,300 ppm/kb, which were not correlated with induction

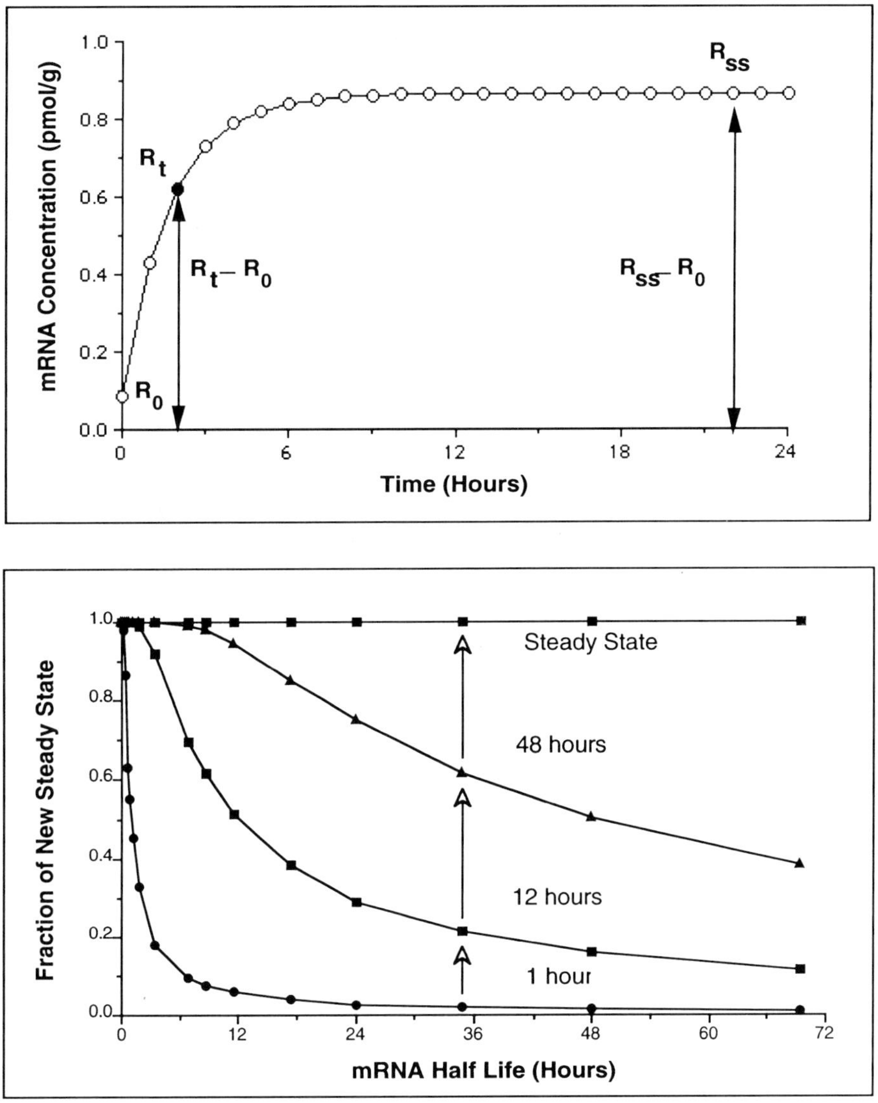

Fig. 2.3. Top: Parameters used in the concept of fractional induction are depicted. The concentrations of mRNA in the initial and final steady states, respectively, are R_0 and R_{ss}. R_t is the value at any time. Bottom: The predicted relationship between a mRNA's chemical half-life and the time course of induction between two steady-state concentrations is summarized, based on the fractional increase predicted from Eq. 2-3.

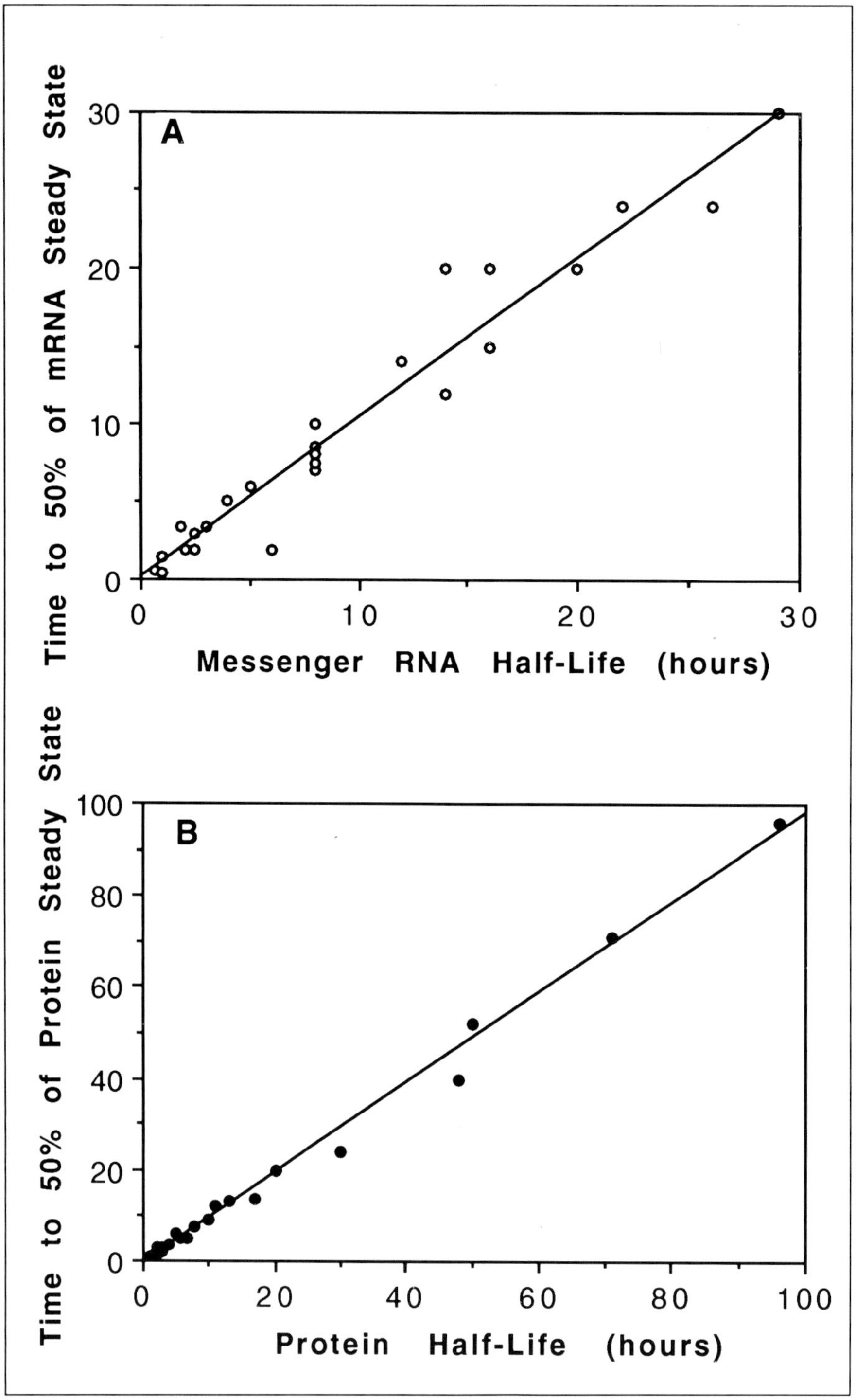

Fig. 2.4. The relationship between half-lives and induction periods for mRNAs (A), and proteins (B) is shown for the gene products listed in Table 1. The time required for mRNAs or proteins to attain 50% of their respective, steady-state concentrations after rates of synthesis changed was used as a criterion. Half-lives and times for 50% induction of messenger RNAs and proteins were obtained from the references listed in Table 2.1; for half-lives that vary, the values that were in effect during the induction periods were used.

Table 2.2. Rates of transcription and degradation for vertebrate mRNAs and proteins

Gene Product	**Rate of Transcription** (parts per million/kb)	**mRNA Half-Life** (hours)	**Protein Half-Life** (hours)	**References**
Albumin	340-797	20	secreted	(24, 25)
Alcohol dehydrogenase	50-320	9-20	72	(26)
Casein	ND-300	5-92	secreted	(27, 28)
Conalbumin	22-320	8	secreted	(29)
Cytochrome P-450IIB1	4-86	3	16	(30, 31)
Cytochrome P-450IA1	5-143	100	36	(32, 33)
Cytochrome P-450IA2	6-94	29	24-72	(32, 33)
Dihydrofolate reductase	1-4	6	97	(34, 35)
Fatty acid synthase	11-116	3	54	(36, 37)
β-Globin	38-324	20	stable	(38, 39)
Glucokinase	2-42	3	12	(40, 41)
Glucose-6-phosphate dehydrogenase	1-5	1-24	15	(42-44)
β-Glucuronidase	8-60	36	84-194	(45-47)
HMG-CoA reductase	5-192	3	3	(48, 49)
Ornithine decarboxylase	11-176	1.7-5.7	0.5	(50, 51)
Malic enzyme	6-25	1.5-11	67	(52-54)
Metallothionein	15-275	4	1-84	(55, 56)
Ovalbumin	1-1210	24	secreted	(5, 6)
PEP carboxykinase	69-1313	0.5-2.5	5	(57, 58)
Pyruvate kinase	0.5-12	1-3	30	(41, 59)
Serine dehydratase	11-59	2	13	(60, 61)
Thymidine kinase	1-4	2.6	2.6	(62, 63)
Tryptophan oxygenase	9-84	1.0	2.5	(62, 64)
Tyrosine aminotransferase	10-110	1.5	2.0	(65-67)
Range of Variation	about 2000-fold	ca. 200-fold	ca. 400-fold	

periods (not shown). Caution in interpreting these data is merited due to the number of technical difficulties in measuring rates of transcription and rates of decay. Nonetheless, theory suggests that the rate of mRNA synthesis would only alter the period needed for induction to a new steady state if it was affected by mRNA concentration. For vertebrate genes, little evidence exists of direct feedback between product levels and transcriptional activity, although potential for this exists.

According to this model, the period required for maximal gene expression depends less on genetic mechanisms than on the physical principle that decay constants govern the time course of change in compartmental systems. If the rates at which specific mRNAs and proteins decay are proportional to product concentrations, this conclusion is inescapable, and the majority of the decay curves that were reviewed in preparing this manuscript were exponential over most of the reported period, indicating apparent first-order decay constants (for example, see refs. 26,28,47,58,62). The data shown in Fig. 2.4 demonstrate that the model of gene expression based on the idea of zero order synthesis and first order decay works surprisingly well despite its limitations. Because the half-lives of mammalian mRNAs and proteins are long in comparison to the other processes involved in gene expression, the time course of enzyme induction is related to the half-lives of the mRNA and protein in question. When a process is affected by the concentration of a protein, or requires that the induced level of product exceed the basal level by a specific amount in order to initiate a response, then the principle of fractional response becomes important. During the course of organic evolution, some genes must have been modified by natural selection to optimize the rate of response or ratio of induced to basal expression.

References

1. Neidhardt FC. Chemical composition of Escherichia coli In: Neidhardt FC ed. Escherichia coli and Salmonella typhimurium. Cellular and Molecular Biology, vol. 1, American Society for Molecular Biology, Washington DC. 1987; 3-6.
2. Boxenbaum H. Time concepts in physics, biology, and pharmacokinetics. J Pharmaceut Sci 1986; 75: 1053-1062.
3. Strehler BL. Time, Cells, and Aging. New York: Academic Press, 1977.
4. Crick F. Central dogma of molecular biology. Nature 1970; 227: 561-563.
5. Palmiter RD. Quantitation of parameters that determine the rate of ovalbumin synthesis. Cell 1975; 4, 189-197.
6. McKnight GS, Palmiter RD. Transcriptional regulation of the ovalbumin and conalbumin genes by steroid hormones in chick oviduct. J Biol Chem 1979; 254, 9050-9058.
7. Jacob F, Monod J. Genetic regulatory mechanisms in the synthesis of proteins. J Mol Biol 1961; 3: 318-356.
8. Pardee AB, Prestidge LS The initial kinetics of enzyme induction. Biochim Biophys Acta 1961; 49: 77-88.
9. Contesse G, Crepin M, Gros F. Transcription of the lactose operon in E. coli. In: Beckwith JR, Zipser D eds. The Lactose Operon. Cold Spring Harbor Laboratory, New York 1970: 111-141.
10. Monod J, Pappenheimer AM, Cohen-Bazire G. La cinetique de la biosynthese de la β-galactosidase chez E. coli consideree comme fonction de la croissance. Biochim Biophys Acta 1952; 9: 648-660.
11. Monod J. The phenomenon of enzymatic adaptation. Growth 1947; 11: 223-289.
12. Knox WE. Two mechanisms which increase in vivo the liver tryptophan peroxidase activity: specific enzyme adaptation and stimulation of the pituitary-adrenal system Br J Exp Path 1951; 32: 462-469.
13. Comings DE, Muhleman D, Dietz GW Jr, Donlon T. Human tryptophan oxygenase localized to 4q31: possible implications for alcoholism and other behavioral disorders. Genomics 1991; 9: 301-308.
14. Comings DE. Blood serotonin and Tourette syndrome. Am J Med Genet 1990; 36: 418-430.
15. Schmid W, Scherer G, Danesch U et al. Isolation and characterization of the rat tryptophan oxygenase gene. EMBO J 1982; 10, 1287-1293.
16. Maezono K, Tashiro K, Nakamura T. Deduced primary structure of rat

tryptophan-2,3-dioxygenase. Biochem Biophys Res Comm 1990; 170: 176-181.
17. Danesch U, Hashimoto S, Renkawitz R, Schutz G. Transcriptional regulation of the tryptophan oxygenase gene in rat liver by glucocorticoids J Biol Chem 1983; 258: 4750-4753.
18. Vom Dahl S, Hallbrucker C, Lang F, Gerok W, Haussinger D. Regulation of liver cell volume and proteolysis by glucagon and insulin. Biochem J 1991; 278: 771-777.
19. Fersht A. In: Enzyme Structure and Mechanism. New York: WH. Freeman and Co, 1985.
20. Moore RE, Goldsworthy TL, Pitot HC. Turnover of 3'-polyadenylate-containing RNA in livers from aged, partially hepatectomized, neonatal, and Morris 5123C hepatoma-bearing rats, Cancer Res 1980; 40: 1449-1457.
21. Hargrove J, Hulsey M, Beale E. The kinetics of mammalian gene expression. BioEssays 1991; 13: 667-674.
22. Hargrove JL, Hulsey MG, Summers AO. From genotype to phenotype: kinetic modeling of gene expression with STELLA II. BioTechniques 1993; 8: 654-659.
23. Arraiano CM, Yancey SD, Kushner SR. Stabilization of discrete mRNA breakdown products in ams pnp rnb multiple mutants of Escherichia coli K-12. J Bacteriol 1988; 170: 4725-4633.
24. Peavy DE, Taylor J. M, Jefferson L. S. Time courses of changes in albumin synthesis and mRNA in diabetic and insulin-treated diabetic rats. Amer J Physiol 1985: 248, E656-E663.
25. Yamauchi A, Imai E, Noguchi T, Tanaka T, Yamamoto S, Mikami H, Fukuhara Y, Fujii M, Orita Y, Kamada T. Albumin gene transcription is enhanced in liver of nephrotic rats. Am J Physiol 1988; 254, E676-679.
26. Felder MR, Watson G, Huff MO, Ceci JD. Mechanism of induction of mouse kidney alcohol dehydrogenase by androgen. J Biol Chem 1988; 263, 14531-14537.
27. Yoshimura M, Oka T. Hormonal induction of β-casein gene expression: Requirement of ongoing protein synthesis for transcription. Endocrinology 1990; 126: 427-433.
28. Guyette WA, Matusik RJ, Rosen JM. Prolactin-mediated transcriptional and post-transcriptional control of casein gene expression. Cell 1979; 17: 1013-1023.
29. McKnight GS, Palmiter RD. Transcriptional regulation of the ovalbumin and conalbumin genes by steroid hormones in chick oviduct. J Biol Chem 1979; 254: 9050-9058.
30. Ravishankar H, Padmanaban G. Turnover of messenger RNA, apoprotein, and haem of cytochrome P450b plus e induced by phenbobarbitone in rat liver. Biochem J 1985; 229: 73-79.
31. Hardwick JB, Gonzalez FJ, Kasper CB. Transcriptional regulation of rat liver epoxide hydrolase, NADPH-cytochrome c reductase, and cytochrome P450b genes by phenobarbital. J Biol Chem 1983; 258: 8081-8085.
32. Steward AR, Wrighton SA, Pasco DS, Fagan, .B Li D, Guzelian PS. Synthesis and degradation of 3-methylcholanthrene inducible cytochromes P450 and their mRNAs in primary monolayer cultures of adult rat hepatocytes. Arch Biochem Biophys 1985; 241: 494-508.

33. Gonzales FJ, Tukey RH, Nebert DW. Structural gene products of the Ah locus. Transcriptional regulation of cytochrome P_1-450 and P_3-450 mRNA levels by 3-methylcholanthrene. Mol Pharmacol 1984; 26: 117-121.
34. Leys EJ, Crouse GF, Kellems RE. Dihydrofolate reductase gene expression in cultured mouse cells is regulated by transcript stabilization in the nucleus. J Cell Biol 1984; 99: 180-187.
35. Rogers S, Wells R, Rechsteiner M. Amino acid sequences common to rapidly degraded proteins: The PEST hypothesis. Science 1986; 234: 364-368.
36. Back DW, Goldman MJ, Fisch JE, Ochs RS, Goodridge AG. The fatty acid synthase gene in avian liver. Two mRNAs are expressed and regulated in parallel by feeding, primarily at the level of transcription. J Biol Chem 1986; 261: 4190-4197.
37. Tweto J, Iberati M, Larrabee AR. Protein turnover and 4'-phosphopantethiene exchange in rat liver fatty acid synthetase. J Biol Chem 1971; 246: 2468-2471.
38. Ganguly S, Skoultchi AI. Absolute rates of globin gene transcription and mRNA formation during differentiation of cultured mouse erythroleukemia cells. J Biol Chem 1985; 260: 12167-12173.
39. Ross J, Sullivan TD. Half-lives of beta and gamma globin messenger RNAs and of protein synthetic capacity in cultured human reticulocytes. Blood 1985; 66: 1149-1154.
40. Iynedjian PB, Gjinovci A, Renold AE. Stimulation by insulin of glucokinase gene transcription in liver of diabetic rats. J Biol Chem 1988; 263: 740-744.
41. Dice JF, Goldberg AL. A statistical analysis of the relationship between degradative rates and molecular weights of proteins. Arch Biochem Biophys 1975; 170: 213-219.
42. Prostko C, Fritz RS, Kletzien RF. Nutritional regulation of hepatic glucose-6-phosphate dehydrogenase. Transient activation of transcription. Biochem J 1989; 258: 295-9.
43. Winberry L, Holten D. Rat liver glucose 6-phosphate dehydrogenase. Dietary regulation of the rate of synthesis. J Biol Chem 1977; 252: 7796-7801.
44. Rudack D, Chisholm EM, Holten D. Rat liver glucose 6-phosphate dehydrogenase. Regulation by carbohydrate diet and insulin. J Biol Chem 1971; 246: 1249-1254.
45. Watson G, Paigen K. mRNA syntheis rates in vivo for androgen inducible sequences in mouse kidney. Mol Cell Biol 1988; 7: 2925-2932.
46. Watson G, Paigen K. Genetic variations in kinetic constants that describe β-glucuronidase mRNA induction in androgen-treated mice. Mol Cell Biol 1987; 7, 1085-1090.
47. Pfister K, Watson G, Chapman V, Paigen K. Kinetics of β-glucuronidase induction by androgens. Genetic variation in the first order rate constants. J Biol Chem 1984; 259, 5816-5820.
48. Clarke C, Fogelman A, Edwards P. Transcriptional regulation of the 3-hydroxy-3-methylglutaryl coenzyme A reductase gene in rat liver. J Biol Chem 1985; 260, 14363-14357.

49. Simonet WS, Ness GC. Turnover rates of HMG CoA reductase mRNA: role of pituitary and thyroid hormones. Biochem Biophys Res Commun 1987; 146: 1033-1039.
50. Laitinen S, Laitinen P, Pajunen P. The effect of testosterone on half-life of ornithine decarboxylase mRNA in mouse kidney. Biochem Int 1984; 9: 45-50.
51. Hayashi S-I, Kameji T, Fujita K, Marakami Y, Kanamoto R, Utsunomiya K, Matsufuji S, Takiguchi M, Mori M, Tatibana M. Molecular mechanism for the regulation of hepatic ornithine decarboxylase. Adv Enz Reg 1985; 23: 311-327.
52. Dozin B, Rall JE, Nikodem VM. Tissue-specific control of rat malic enzyme activity and messenger RNA levels by a high carbohydrate diet. Proc Nat Acad Sci USA 1986; 83: 4705-4709.
53. Goldman MJ, Back DW, Goodridge AG. Nutritional regulation of the synthesis and degradation of malic enzyme messenger RNA in duck liver. J Biol Chem 1985; 260: 4404-4408.
54. Back DW, Wilson SB, Morris SM, Goodridge AG. Hormonal regulation of lipogenic enzymes in chick embryo hepatocytes in culturel. J Biol Chem 1986; 261: 12555-12561.
55. Durnam DM, Palmiter RD. Transcriptional regulation of the mouse metallothionein-I gene by heavy metals. J Biol Chem 1981; 256: 5712-5716.
56. Ohi S, Cardenosa G, Pine R, Huang PC. Cadmium-induced accumulation of metallothionein messenger RNA in rat liver. J Biol Chem 1981; 256: 2180-2184.
57. Lamers N, Hanson R, Meisner H. cAMP stimulates transcription of the gene for cytosolic phosphoenolpyruvate carboxykinase in rat liver nuclei. Proc Nat Acad Sci USA 1979; 79: 5137-5142.
58. Hod Y, Hanson R. Cyclic AMP stabilizes the messenger RNA for phosphoenolpyruvate carboxykinase GTP; against degradation. J Biol Chem.1988; 263: 7747-7752.
59. Vaulont S, Munnich A, Decaux JF, Kahn A. Transcriptional and post-transcriptional regulation of L-type pyruvate kinase gene expression in rat liver. J Biol Chem 1986; 261: 7621-7625.
60. Noda C, Tomomura M, Nakamura T, Ichihara A. Hormonal control of serine dehydratase mRNA in primary cultures of hepatocytes. J Biochem Tokyo 1984; 95:37-45.
61. Noda C, Yakiyama M. Nakamura T, Ichihara A. Requirements of both glucocorticoids and glucagon as co-inducers for activation of transcription of the serine dehydratase gene in cultured rat hepatocytes. J Biol Chem 1988; 263: 14764-14768.
62. Schimke RT. Control of enzyme levels in mammalian tissues. Adv Enzymol 1973; 37: 135-187.
63. Coppock DL, Pardee AB. Control of thymidine kinase mRNA during the cell cycle. Mol Cell Biol 1987; 7: 2925-2932.
64. Danesch U, Hashimoto S, Renkawitz R, Schutz G. Transcriptional regulation of the tryptophan hydroxylase gene in rat liver by glucocorticoids. J Biol Chem 1983; 258: 4750-4753.

65. Lee K.-L, Reel JR, Kenney FT. Regulation of tyrosine α-ketoglutarate transaminase in rat liver. IX. Studies on the mechanism of hormonal inductions in cultured hepatoma cells. J Biol Chem 1970; 245: 5806-5812.
66. Kenney FT. Induction of tyrosine α-ketoglutarate transaminase in rat liver. IV. Evidence for an increase in the rate of enzyme synthesis. J Biol Chem 1962; 237: 3495-3498.
67. Hashimoto S, Schmid W, Schutz G. Transcriptional activation of the rat liver tyrosine aminotransferase gene by cAMP. Proc Nat Acad Sci USA 1984; 81: 6637-6641.

CHAPTER 3

Modeling Gene Expression with Simulation Software

Although it may seem a paradox, all exact science is dominated by the idea of approximation. When a man tells you that he knows the exact truth about anything, you are safe in inferring that he is an inexact man.
—Bertrand Russell

USING THE IDEA OF APPROXIMATION

The traditional approach to the mathematics of dynamic systems involves writing and solving differential equations that predict behavior of the system at any future moment. This approach represents the highest standard of kinetic modeling, but few biologists have sufficient familiarity with differential equations to employ it. Engaging a colleague to derive the equations and write computer programs is helpful, but does not enhance thinking by providing immediate feedback, and the biologist does not have the flexibility to change the equations or the program. In view of this, it is reasonable to ask how often biological systems meet the rigid set of conditions required for analytical differential equations to predict a correct solution. In the case of gene expression, it is doubtful that a true steady state exists, and no stimulus acts instantaneously because the system of interest is usually a cell population that is affected by stimuli that must diffuse into individual cells and recruit *trans*-acting factors to specific genes before a new rate

---

STELLA®II is a registered trademark of High Performance Systems, Inc. New Lyme, CT 03824, telephone 603-862-2186; the Dynamo Plus simulation program for MS-DOS systems is available from Pugh-Roberts Associates, 41 William Linsky Way, Cambridge, MA 02142. Macintosh is a trademark of Apple Computer, Inc.

of transcription begins. Furthermore, biological assays are estimates of population means, and generally experimental data vary by 5-10% around the true mean; this is why statistical methods are used to test for significance of observations. These arguments suggest that most biologists would be justified in accepting the idea of approximation if it simplifies the task of predicting system behavior and provides rapid feedback.

The examples provided below demonstrate an alternative to using standard differential equations, which does not require intervention by a mathematician, and is supported by new computer software for dynamic modeling. If the idea of approximation within specified limits is acceptable, then finite difference equations that are constructed interactively at a computer workstation can be used to predict system dynamics; this approach was taken by Jay Forrester and colleagues at MIT who devised a numerical technique of computer simulation for the purpose of analyzing the interactions between resource limitation and growth of the human population.[1,2] Basically, the technique is a form of mathematical modeling based on a simulation program that permits the user to convert a diagram of a system into a functioning program. The original mainframe computer program that was devised for this task was called Dynamo, and a newer program based on these ideas is called STELLA®II.[3] This chapter will compare results obtained using a computer program that solves two differential equations by the traditional method with results obtained with finite difference equations using the STELLA II package.

AN EXAMPLE OF THE TRADITIONAL APPROACH

The simplest kinetic model of gene expression based on the assumptions of zero-order synthesis and first-order decay[4-6] can be represented as a pool of translationally active mRNA that encodes a specific protein (Fig. 3.1A). In this model, the rate of mRNA synthesis (λ_1) is equivalent to the rate at which active mRNA appears in the cytoplasm, and represents an aggregate rate of transcription, processing, and nucleocytoplasmic transport. The rate of protein synthesis (λ_3) is assumed to be linearly proportional to mRNA concentration, and the protein and mRNA are degraded with first-order kinetics (λ_2 and λ_4, respectively). From these assumptions, the following set of equations may be derived, in which R indicates the concentration of a specific mRNA, and P designates the concentration of the protein encoded by that mRNA. R_0 is defined as the concentration of mRNA in the system during the initial steady-state, R_t the amount at any moment between initial and final steady states, and R_{ss} the concentration at the new steady-state; protein concentrations at different times are designated by the same subscripts. The rate parameters for synthesis and degradation are indicated by the letter, λ, with subscripts as shown in Figure 3.1.

Equation 1: The rate of change in concentration of a gene product between two, steady state levels is a function of its rates of synthesis and decay, assuming zero-order synthesis and first-order decay:

$$\frac{dR}{dt} = \lambda_1 - \lambda_2 R_t \qquad \text{Eq.3.2}$$

Equation 2: The solution to Eq. 3.1 shows that the time course required for the concentration of a gene product to attain a new steady-state is related to the decay constant or half-life, no matter whether the concentration is increasing or decreasing:

$$R_t = R_0 + (R_{ss} - R_0)(1 - e^{-\lambda_2 t}) \qquad \text{Eq. 3.2}$$

Equation 3: In the initial and final steady-states, the rate of synthesis equals the rate of decay (which is the product of the rate constant for decay and the concentration of product). The initial and final concentrations of mRNA equal the ratio between the two values in effect during the initial and final steady states:

$$\lambda_1 = \lambda_2 R_{ss}, \text{ or } R_{ss} = \frac{\lambda_1}{\lambda_2} \qquad \text{Eq. 3.3}$$

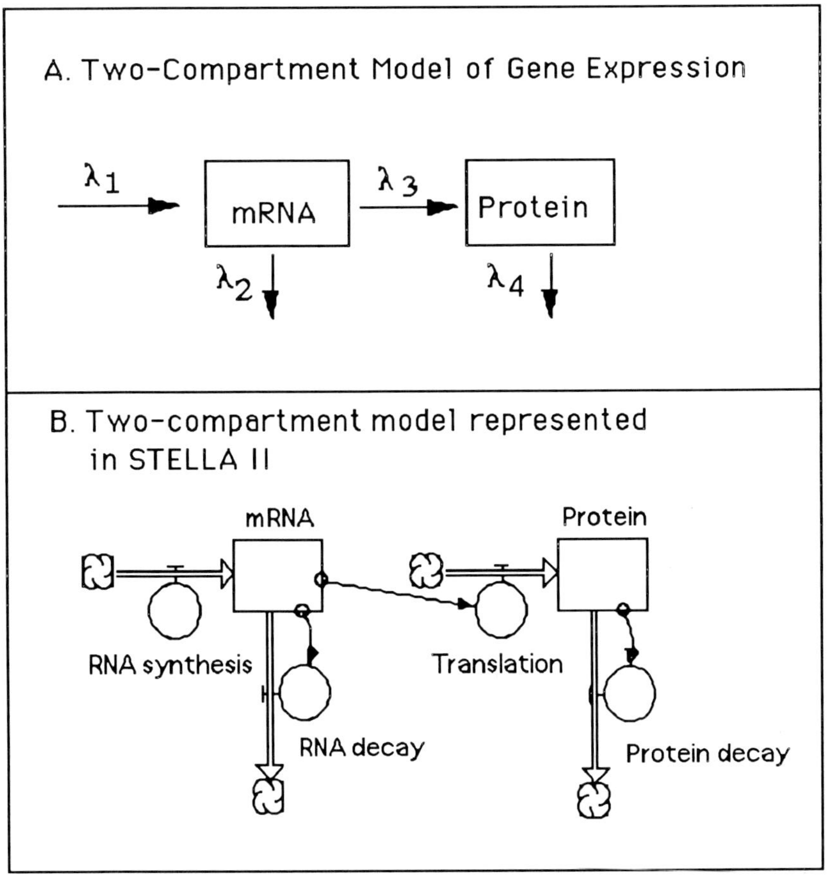

Fig. 3.1. A. Traditional depiction of a two-compartment model of gene expression, with rates of synthesis and decay indicated by arrows flowing into and out of the compartments. B. The same model dipicted with the user interface for the STELLA II simulation program. The arrows shown in the upper diagram are replaced by icons called flows, and proportionality, such as first order decay, is represented by arrows called connectors. Each circle and rectangle is associated with a dialog box for entering numbers or equations.

An equation may be written to show the relationship of protein concentration to mRNA concentration using the assumptions that protein synthesis is related to mRNA concentration by a first-order rate constant for translation, and that protein decay follows first-order kinetics. In this solution, the rate constants for synthesis and decay of mRNA are designated λ_1 and λ_2, respectively, and the corresponding values for protein are designated λ_3 and λ_4:

Equation 4: The concentration of a protein may be expressed in terms of the concentration of the corresponding mRNA:

$$P_t = P_0 + (P_{ss}-P_0)(1-e^{-\lambda_4 t}) - \lambda_3(R_{ss}-R_0)\left\{\frac{e^{-\lambda_4} - e^{-\lambda_2 t}}{\lambda_2 - \lambda_4}\right\} \quad \text{Eq. 3.4}$$

Note that the first two terms on the right side of equation 3.4 are identical in form to the solution to the single-compartmental model (Eq. 3.2). The last term on the right introduces a delay between induction of mRNA and induction of the corresponding protein.

A traditional computer program based on the two-compartment model of gene expression solves equations 3.2 and 3.4 when initial values for the parameters are supplied, along with one or more changes from the original steady state.[7] The parameters needed are readily calculated from experimentally determined concentrations and half-lives, as shown in Table 3.1. No matter which programming language was

Table 3.1. Example of calculations for kinetic parameters and steady-state values when modeling gene expression

	Rate of Synthesis (units/h)	**Half-time of Decay** (h)	**Rate Constant** (h^{-1})	**Half-time of Transit** (h)	**Rate Constant** (h^{-1})	**Concentration**
Primary Transcript	1.0[a]	1.0	0.693[b]	0.16	4.15	0.21
Nuclear mRNA	0.87[c]	0.33	2.08	0.33	2.08	0.21
Cytoplasmic mRNA	0.44	2.0	0.347	N.A.[d]	N.A.	1.26
Protein	126[e]	4.0	0.173	N.A.	N.A.	728

[a]The rate of synthesis in this example was assumed to be 1 unit per hour; units could be molecules·($cell^{-1} \cdot h^{-1}$) or moles·(grams of $tissue^{-1} \cdot h^{-1}$).

[b]The kinetic parameter equals 0.693/half-life.

[c]The rate of synthesis at steady state for each intermediate equals the rate constant for conversion of the preceding constituent times the concentration of that constituent.

[d]N.A., not applicable.

[e]In this example, the rate of protein synthesis was arbitrarily set at 100 moles of protein per mole of mRNA per hour, which is low compared to rates in tissues such as liver.

p. 36—Numerator in last term in Eq. 3.4 is lacking a t in the exponent. It should read: $[e^{-\lambda_4 t} - e^{-\lambda_2 t}]$ not $[e^{-\lambda_4} - e^{-\lambda_2 t}]$

chosen for the work, verifying the equations and writing the program would require several hours, and any changes would necessitate extensive rewriting of the code to accommodate new equations of progressively greater complexity. As more compartments were added to include items such as translational control of protein synthesis or nuclear mRNA dynamics, the equations and program would rapidly become complex, and the time required to carry out the work would grow. Now compare this method with the mode of operation and results of a program designed for computer simulation using compartmental analysis.

SOLVING A MODEL OF GENE EXPRESSION WITH A SIMULATION PROGRAM

STELLA II is basically a computer program for analyzing system dynamics using compartmental analysis, and most other programs for mathematical modeling have similar capabilities. The user interface in the STELLA II program is a blank screen on which a diagram may be drawn using icons from the tool kit, just as is done with most programs for computer graphics (Fig. 3.2). The difference is that the completed diagram is both a visual model of the system and also includes a set of equations which the program solves using numerical methods; the user may view the equations and print them to keep a record, but this is not necessary, and the equations are hidden during normal operation. The program functions by adding the material that flows into each compartment during each specified interval to the pre-existing

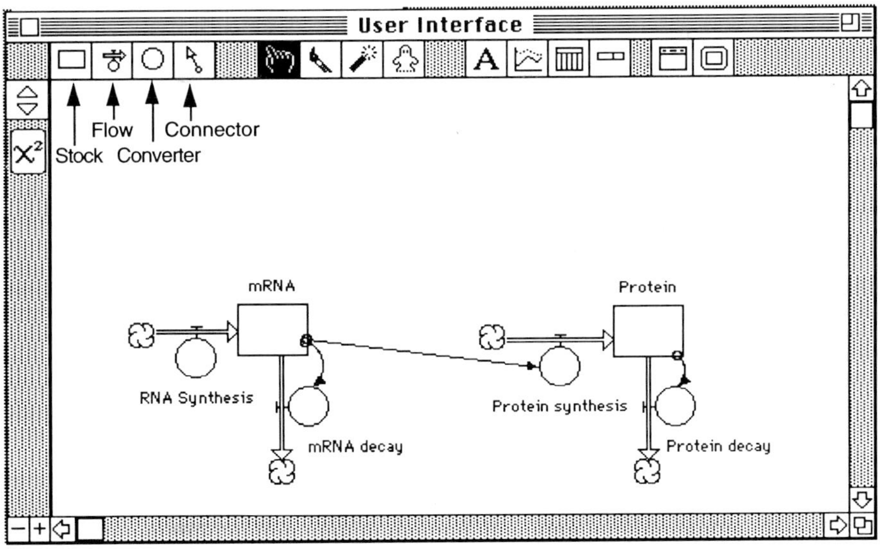

Fig. 3.2. The user interface for model construction with the STELLA II simulation program is shown. Models are constructed using the four building blocks indicated along the bar at the top, and other tools are used to customize the diagram. The two symbols to the right of the letter A activate a graphical display and a tablular display that are used to view output. Equations associated with the model may be viewed by clicking the symbol labeled c² on the left side of the diagram.

material, and subtracting the material that moves out of each compartment during the same period. It repeats this process until the period specified for the simulation is complete, and prints the results as a graph or table. As the calculation interval is made smaller, the results approach the same limiting value that would be obtained with an analytical solution to a differential equation.

In order to compare results of the traditional approach with those obtained by the method of approximation, the block diagram of gene expression represented in Fig. 3.1A was converted to a computer program using STELLA II.[8] The program uses four icons to assemble system models; these units are depicted in the "tool box" on the upper left side of Fig. 3.2. *Stocks,* which represent compartments or pools in which gene products may accumulate, are depicted as open rectangles. To add a stock to a model, the icon is selected with the mouse by clicking on the first rectangle in the tool box, moving it to its destination, and depositing it with a second click of the mouse button. In the model shown here, material in the pools has units of mass or concentration, such as picomoles per cell. *Flows* of material into and out of the reservoirs are represented as pipes that contain a source, a valve, and a direction (arrowhead; second building block in tool kit). Flows usually represent physical transfer, and have units of mass or concentration transferred per unit of time. If material flows into or out of the system from an external source that does not affect the behavior of the system, the source is represented by a "cloud." Feedback relationships or proportionalities among components are indicated by arrows called *connectors,* and the nature of the relationship is specified using building blocks called *converters* (open circles), which may function as regulators or may be used to keep units consistent among different parts of a system. Each flow is regulated by at least one converter. The other tools in the margin adjacent to the building blocks are used to select or remove elements, to configure the model according to the goals of the designer, and to display the results of simulation.

After a systems diagram has been prepared with STELLA II, equations are constructed that give numerical solutions for the quantitative relationships indicated by the diagram (Fig. 3.3). This is done interactively by opening *dialog boxes* (Fig. 3.4) for each building block that contains a question mark. For example, the initial RNA concentration at steady state equals the rate of synthesis divided by the rate constant for decay, and this value may be calculated and typed into the highlighted line. Equations may be entered into the box either by typing in the needed information, or by selecting appropriate functions from an extensive library of mathematical functions called the "Built-ins" list. Alternatively, input may be derived from a graph or spreadsheet by clicking a button at the bottom of the dialog box. The equations that are generated during model construction may be viewed and modified by selecting "Equations" from the Windows menu. Figure 3.3 shows the equations that correspond to the kinetic model of Figure 3.1 with one set of initial conditions.

STELLA II generates finite difference equations that provide solutions at short intervals (dt or "DT"). The calculation interval has a

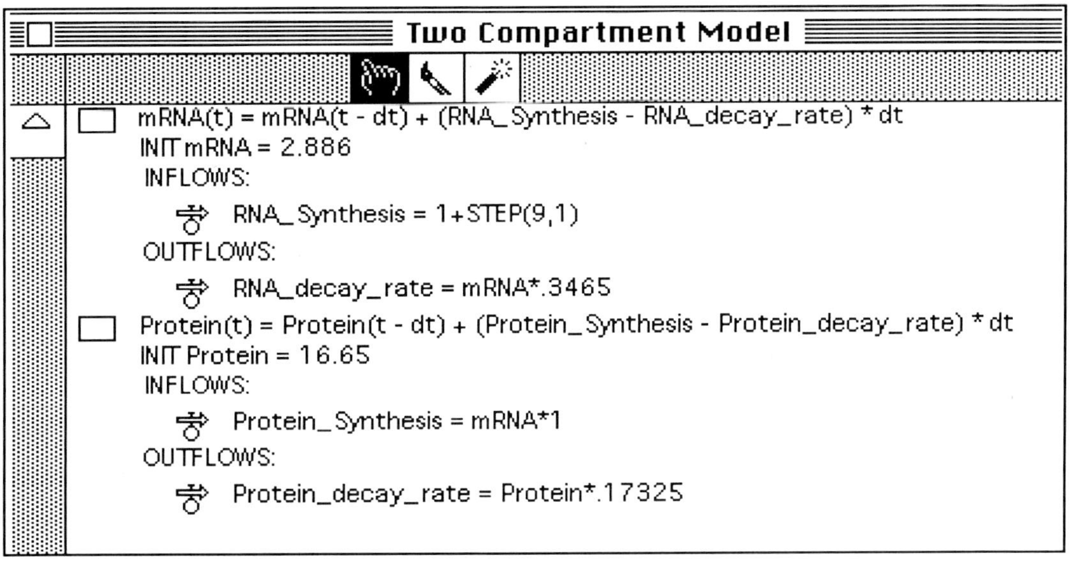

Fig. 3.3. Finite difference equations corresponding to the two-compartment model are shown for comparison to the traditional differential equations shown in the text. The equations are written interactively as the diagram is constructed, and are solved by numerical algorithms.

Fig. 3.4. Each open symbol shown in the user interface is associated with a dialog box. This one is linked to the icon for mRNA synthesis, and indicates that mRNA synthesis will initially have a value of 1 unit per hour, but will increase (step) to a value of 10 at 1 hour. Any function from the "Built-ins" list may be selected, or an equation may be typed into the area under "RNA synthesis =..."

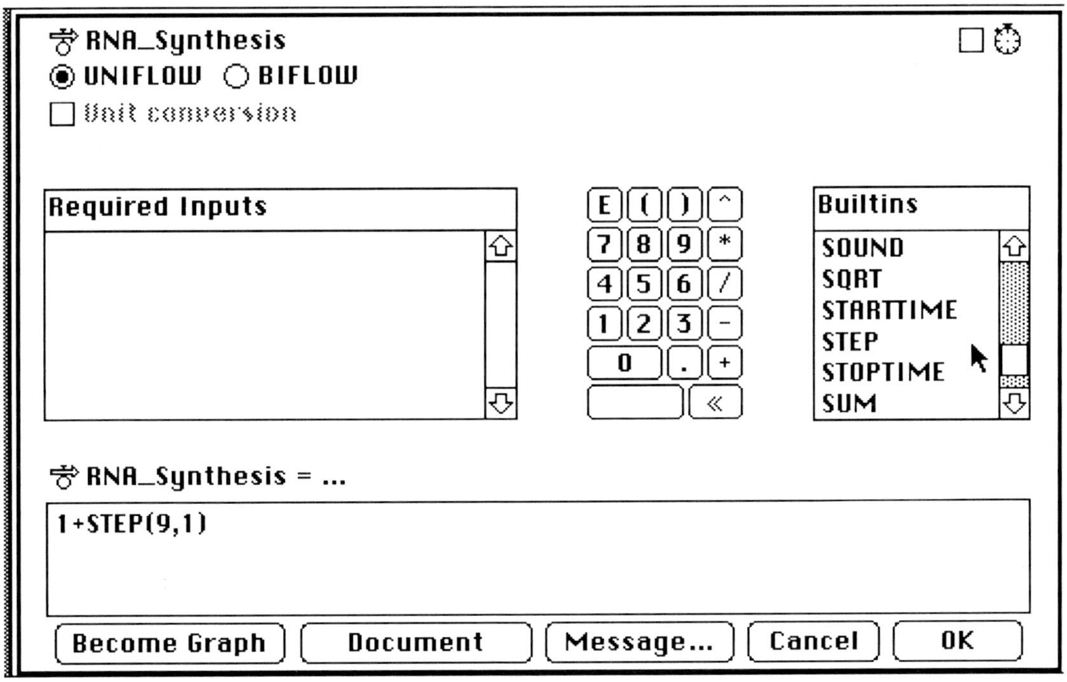

default value of 0.25 time units but this may be changed to any value between zero and one. STELLA solves the equations using one of three numerical methods for estimating values for each interval. The results converge with exact solutions when the calculation interval is short; as a rule, the calculation interval should not exceed one-half of the shortest time delay or time constant in the model. It is very important for users to understand the potential pitfalls of inexact solutions and test the effects of changing the calculation intervals.

The model of gene expression shown in Figure 3.1 is based on the assumptions that the rate of mRNA synthesis is zero-order (i.e., equal amounts are produced in each interval), whereas protein synthesis and decay processes are first-order (i.e., proportional to concentration of mRNA or protein). To indicate that flows are proportional to existing mRNA or protein using STELLA, a connector is used to join the stock to the valve on the outflow (Fig. 3.2). Because most proteins and mRNAs decay in this fashion, *the time scale of accumulation for mRNAs and proteins is thought to be determined largely by their rates of decay*. According to the kinetic model of Figure 3.1, the major reason for the delay between initiation of transcription and accumulation of protein is that a new steady state in product concentration cannot be attained until five half-lives have elapsed. To duplicate this feature, decay constants were defined as follows:

$$\lambda = \ln 2/\text{half-life} = 0.693/\text{half-life} \qquad \text{Eq. 3.5}$$

For example, a half-life of 1 hour is equivalent to a decay constant of 0.693 h^{-1}. As an aid to setting the scale in graphs generated with STELLA, it is useful to calculate the initial and final steady-state values of mRNA (R_{ss}) and protein (P_{ss}). Notice from Figure 3.1 that the quantity of material in any compartment at steady state equals the rate of inflow divided by the sum of all applicable rate parameters for outflow. In the model shown in Figure 3.1, each compartment has only one outflow, and Eq. 3.3 gives the steady-state concentration of mRNA. The corresponding relationship for the protein is:

$$P_{ss} = R_{ss}\,(\lambda_3/\lambda_4) \qquad \text{Eq. 3.6}$$

where λ_3 is the amount of protein synthesized per molecule of mRNA per unit of time and cell mass, and λ_4 is the rate of protein degradation calculated in the same manner as shown in Eq. 3.5.

To simulate gene expression, the kinetic model shown in Figure 3.2 only requires that four kinetic parameters be known, including the rate of mRNA synthesis, the mRNA's half-life and rate of translation, and the protein's half-life. In order to verify that STELLA would produce the same results as our prior program, values were compared for the same kinetic constants and calculation intervals (0.25 hour and 0.05 hour). When the shorter calculation interval was used, the maximum difference among values compared to results from the analytical solution equaled 0.05%, or five parts per thousand for the hypothetical mRNA (Fig. 3.5A) and the protein (Fig. 3.5B). It is essential that users of this program test the effects of changing the calculation intervals to ensure that minimum values have been achieved.

The time required to construct the model in STELLA II and perform the simulation was 6 minutes; a few more minutes was needed

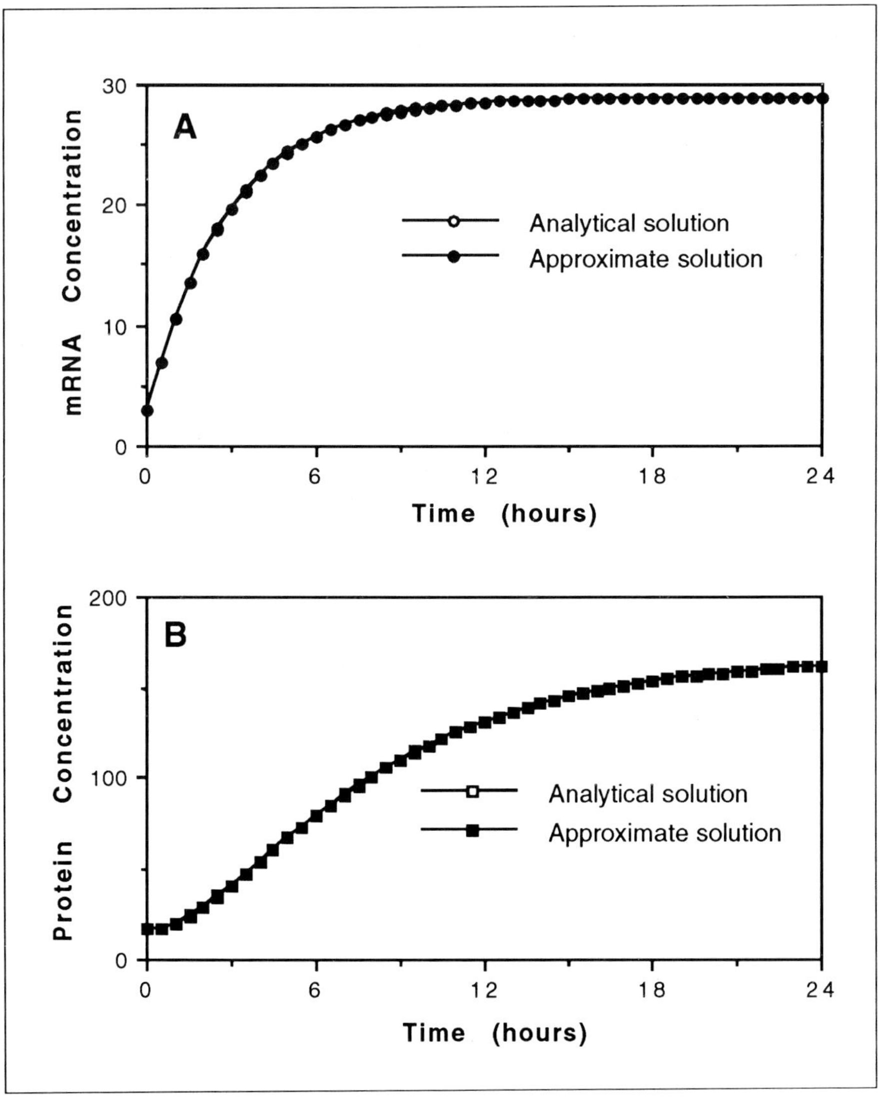

Fig. 3.5. Results of simulations are compared for a program based on an analytical solution to the two-compartment model and the approximate solutionn obtained with the STELLA II package. The largest difference observed when the calculation interval equalled 0.025 h was 0.5% of the analytical solution, which cannot be detected when the results are plotted together.

to export the data to a graphics package so that results of the two simulations could be compared. The first time a colleague was asked to write a program in the traditional manner, the time required to complete the task was several weeks. Considering the degree of variability observed for most biological data, the new simulation program plainly offers several advantages for the rapid design and testing of predictive models.

SIMULATION PROGRAMS MAKE MATHEMATICS ACCESSIBLE TO BIOLOGISTS

Any model that attempts to account for all the known points of control in gene expression will produce extremely complex equations when the traditional approach is taken, and it is unlikely that many biologists will choose this approach. For this reason, new computer programs that enable non-mathematicians to set up and solve equations using a visually-based interface offer a splendid opportunity. The new programs enable any scientist to employ compartmental analysis of gene expression, with solutions expressed as finite difference equations, as an alternative to analytical solutions to differential equations solved in the traditional manner. Importantly, the degree of error introduced by this method can be kept in the same range as the error in the experimental measurements. The investigator must recognize that programs such as STELLA II provide easy access and rapid feedback, but yield solutions that are slightly different from analytical equations. Each user must exercise care in constructing the models, so that modeling can be used for its valid purpose of understanding system dynamics. The method of approximation has a valid role in analyzing system dynamics, but specially trained colleagues should be consulted when a higher degree of rigor is required, or a question arises about proper construction of the model. Despite the advantages of finite difference equations for compartmental analysis, analytical differential equations remain the standard for most professional journals. Investigators who prefer to use analytical equations may employ other simulation programs of moderately greater difficulty (cf. chapter 1). Each of these programs has a user interface that permits simulations to be performed in much shorter time than was required for manual programming, and most also provide features such as parameter estimation and curve-fitting.

REFERENCES

1. Forrester JW. World Dynamics. Cambridge MA, Wright-Allen Press, 1971.
2. Meadows DH, Meadows DL, Randers J, and Behrens WW. The Limits of Growth: A report for the club of Rome's project on the predicament of mankind. New York: Universe Books, 1972.
3. Richmond B, Peterson S. STELLA II An Introduction to Systems Thinking. Hanover, New Hampshire: High Performance Systems, Inc, 1992.
4. Berlin CM, Schimke RT. Influence of turnover rates on responses of enzymes to cortisone. Mol Pharmacol 1965; 1:149-156.

5. Schimke RT. Control of enzyme levels in mammalian tissues. Adv Enzymol 1973; 37: 135-187.
6. Yagil G. Quantitative aspects of protein induction. Curr Top Cell Reg 1975; 9: 183-286.
7. Hargrove JL, Schmidt FH. The role of mRNA and protein stability in gene expression. FASEB J 1989; 3: 2360-2370.
8. Hargrove JL, Hulsey MG, Summers AO. From genotype to phenotype: kinetic modeling of gene expression with STELLA II. BioTechniques 1993; 8: 654-659.

CHAPTER 4

Kinetic Modeling of Gene Expression

Lwoff, after considering this strange result for a moment, said to me, "That could have something to do with enzyme adaptation." "Enzyme adaptation? Never heard of it!" I said.

—Jacques Monod

TOWARDS A COMPREHENSIVE MODEL OF GENE EXPRESSION

The idea that rates of elimination influence the time course of approach to new, steady-state concentrations has been validated for enzymes and mRNAs, but the two-compartment model is incomplete. It does not account for the processes by which transcription is activated, which frequently involve the regulation of DNA-binding proteins by ligands or by covalent modification. Neither does it include the processes involved in mRNA processing and transport, or the impact of the completed protein on cellular metabolism. To be useful to the molecular biologist, a simulation tool should be able to describe any intermediate that could potentially affect the outcome of an experiment, and this requires information about each rate of formation, transfer, or elimination. In the sequential reactions of gene expression, the processing or transfer of each intermediate not only affects the yield, but also introduces a delay that is proportional to the half-time of each step. Many molecular biologists need to make predictions concerning the effects of altered transcription, nuclear mRNA metabolism, and translational or post-translational controls. Although each reaction in the process of gene expression is characterized by an individual rate or half-time for completion, the system cannot attain equilibrium until each process has attained a constant rate. The interval required for the system to equilibrate may be defined as the *relaxation time,*[1,2] and experimental data that deal with induced mRNA or protein synthesis reflect this aggregate property. This chapter addresses this concept, shows how to go about assembling a basic model, and demonstrates how kinetic modeling may be used to make predictions and integrate primary data concerning more inclusive models. The conclusions derived here are based on simulations performed with STELLA II software, which was discussed in Chapter 3, and the presentation assumes that the reader is familiar with the symbols employed by this computer interface.

CREATING A COMPUTER PROGRAM TO SIMULATE GENE EXPRESSION

A general model of gene expression should include at least three components, (1) the phase of transcriptional activation, which is a non-linear process; (2) the central steps of nuclear mRNA processing, cytoplasmic mRNA translation, and protein accumulation, and (3) functional effects of the protein, which are non-linear for enzymes. The first step in compartmental analysis is to create a diagram of the system of interest, and Figure 4.1 illustrates a schema that begins with transcription and includes pools of nuclear mRNA, cytoplasmic mRNA, and protein. This is the core of the model, and will be described first. Precedents for modeling the activation of transcription, translational controls, and the impact of proteins on metabolic control will be discussed at the end of the chapter.

As represented in the STELLA II modeling system, each constituent that accumulates is represented as a rectangle (reservoir or stock), and each process that generates the constituent is represented by a flow into the reservoir, with the fractional transfer rates designated λ_i. An investigator who wished to include other features would follow the same procedure, but would add extra compartments, such as a non-

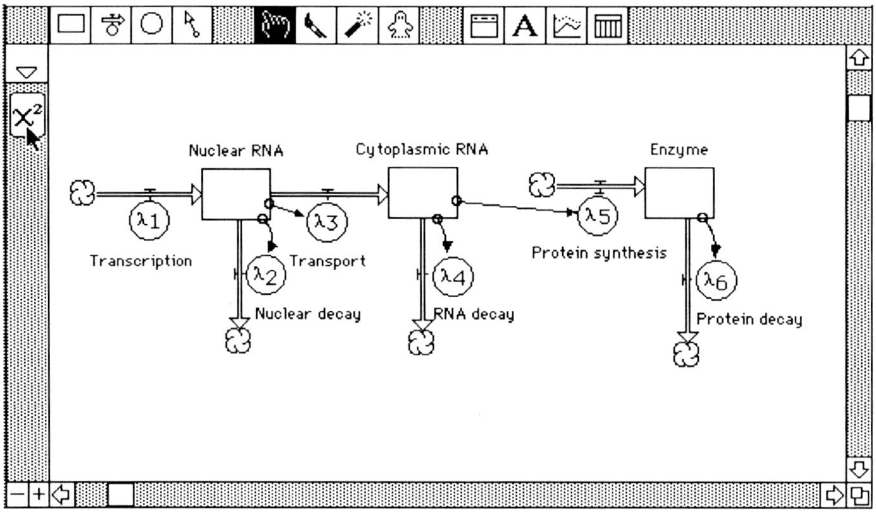

Fig. 4.1. A kinetic model of eukaryotic gene expression including nuclear decay and nucleocytoplasmic transport was produced using STELLA®II. Rates of synthesis and decay or interconversion were entered by opening dialog boxes associated with each open symbol in the diagram, and the program automatically builds a set of finite difference equations as described in chapter 3. Simulations performed with this model predict a rapid accumulation of nuclear mRNA, followed by sigmoidal increases in cytpoplasmic mRNA and protein.

translated pool of mRNA, or representations of protein processing or secretion. When the diagram is completed, the modeler's next task is to define initial conditions and any functions that indicate proportional relationships. The present model will be constructed on the assumption that rates of transfer or degradation are linear, first-order functions, which can be represented by constant fractional transfer. The STELLA II program requests these values by showing a question mark in each icon that has not been defined. The modeler must then open dialog boxes associated with each icon and type in a value.

The simplest way to define initial values for a kinetic model is to write equations to balance the flow of material into and out of each compartment. When a system attains equilibrium the net movement of material into and out of each compartment is balanced, and the concentration in each compartment equals the sum of all rates of inflow divided by the sum of all rates of outflow. When flow out of one compartment into another is proportional to concentration, the rate of formation of each product is equal to the concentration of the previous intermediate multiplied times the fractional amount that enters the next compartment (C_{i-1}times λ_{i-1}). Similarly, each outflow equals the concentration in the pool of interest times the appropriate transfer rate (C_i times λ_i). To show what this means, suppose the nuclear mRNA is designated N, the cytoplasmic mRNA is R, and the protein is called P (Fig. 4.1). Material in each of these compartments is predicted to accumulate and decay according to the elimination rates associated with that compartment, with the net effect that each compartment introduces a delay in the flow of genetic information (Fig. 4.2). The zero-order rate of formation of nuclear mRNA equals λ_1, the rate of transcription; the rate of formation of cytoplasmic mRNA equals λ_3N, and the rate of formation of protein equals λ_5R. Based on Fig. 4.1, the following equations can be written to express the concentration of each product at steady-state:[3]

$$N= \lambda_1/(\lambda_2 + \lambda_3) \qquad \text{Eq. 4.1}$$

$$R= \lambda_3N/\lambda_4 = [\lambda_1\lambda_3]/[(\lambda_2 + \lambda_3)(\lambda_4)] \qquad \text{Eq. 4.2}$$

$$P= \lambda_5R/\lambda_6 = [\lambda_1\lambda_3\lambda_5]/[(\lambda_2 + \lambda_3)(\lambda_4\ \lambda_6)] \qquad \text{Eq. 4.3}$$

Although the rate of transcription, λ_1, is assumed to be constant under the initial and final conditions of the experiment, it is a function of the number of active genes in the cell population of interest, and the average rate of transcription from each. If the investigator desired, this term could be expressed as a function of cell number and the concentration of agents that affect transcription. According to the diagram, the newly synthesized mRNA enters a nuclear pool, and is either degraded (rate=λ_2 h^{-1}), or is processed and transported to the cytoplasm (rate=λ_3 h^{-1}). If the rate of production equals 1 molecule per cell per hour, and the half-time for each transfer is 42 minutes (0.693 hour), then the fractional transfer rates equal 0.693/0.693, or

1 h^{-1}, and the concentration of nuclear mRNA equals 1/(1+1), or 0.5 molecules $cell^{-1}$ h^{-1}. The rate at which mRNA will appear in the cytoplasm equals the concentration times transfer rate, or 0.5 times 1.0, so that the rate of appearance in the cytoplasm equals 0.5 molecules h^{-1}. To calculate mRNA concentration at steady state, the half-life must be known. If the half-life equals 2 h, then the fractional rate of disappearance is 0.693/2, or 0.3465 h^{-1}. This means that the concentration of mRNA at steady state will equal 0.5 molecules per cell per hour divided by 0.3465 h^{-1}, or 1.44 molecules per cell. Similarly, if 100 molecules of protein are made per molecule of mRNA per hour, and the protein has a half-life of 4 h, then the fractional turnover rate of the protein is 0.693/4 h, or 0.1733 h^{-1}, and the protein concentration equals 1.44(100/0.1733), or 831 molecules per cell.

An important conclusion may be derived from the steady-state equations. The equations predict that only changes in the rate of transcription (λ_1) and the rate of decay of cytoplasmic mRNA (λ_4) would produce equivalent effects on mRNA concentration at the steady state. Changes in the terms for nuclear decay or nucleocytoplasmic transport will produce smaller effects because these two processes compete for the same nuclear precursor; if the model included an extra term that represented processing of the primary transcript, the same conclusion would apply. The competition is expressed quantitatively by representing these rates as additive terms in the denominator of Eqs. 4.2 and 4.3. An increase in nucleocytoplasmic transport without an increase in transcription would deplete the pool of nuclear mRNA,

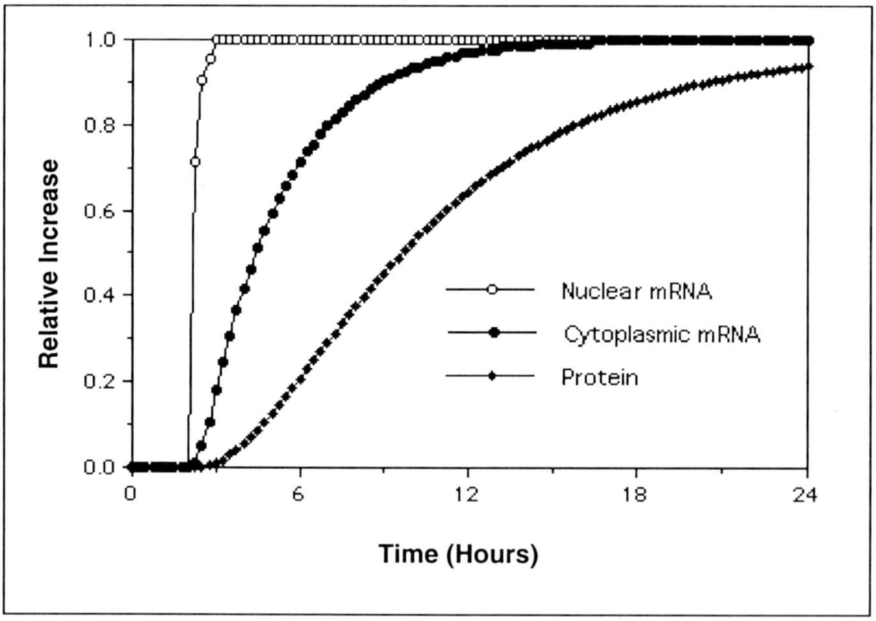

Fig. 4.2. Predicted time course for induction of nuclear precursors, mRNA, and protein after the rate of transcription increases. The kinetic model shown in Fig. 4.1 was used with rates of decay corresponding to the following half-lives: nuclear mRNA, 20 min; cytoplasmic mRNA, 2 h; protein, 4 h.

and therefore would not proportionally increase the amount of cytoplasmic mRNA. For instance, if each parameter had a value of 1, then cytoplasmic mRNA (R) would equal 0.5, and a tenfold increase in rate of transcription or decrease in decay of cytoplasmic mRNA would produce a tenfold increase in RNA. A hand-held calculator can be used to show that a tenfold increase in the rate of processing or export, or increased stabilization of nuclear precursor, will increase the concentration by less than twofold (from 0.5 to 0.9). A larger effect would be observed, however, in the unusual case in which the majority of the transcript was normally degraded in the nucleus. Then nuclear stabilization would increase efflux to a greater extent than in the example just given; only if the cell could choose between complete degradation and no degradation would the effect be equivalent to a corresponding change in transcription. This suggests that control mechanisms that enhance transcription, stabilize cytoplasmic mRNA, or control translation are all more efficient than mechanisms that act on processing or nucleocytoplasmic transport of mature mRNA.

A change in the half-life of the cytoplasmic mRNA would be expected to produce a large effect on the time course of induction because λ_4 represents the time constant for change in concentration of translated mRNA. However, the time scale of change for mRNA precursor is expected to be short, as dictated by $\lambda_2 + \lambda_3$. For most eukaryotic mRNAs, the rate of cytoplasmic decay is much slower than the time constants for the nuclear processes, and the change in concentration of mRNA in the cytoplasm would be expected to be influenced relatively little by the nuclear events.

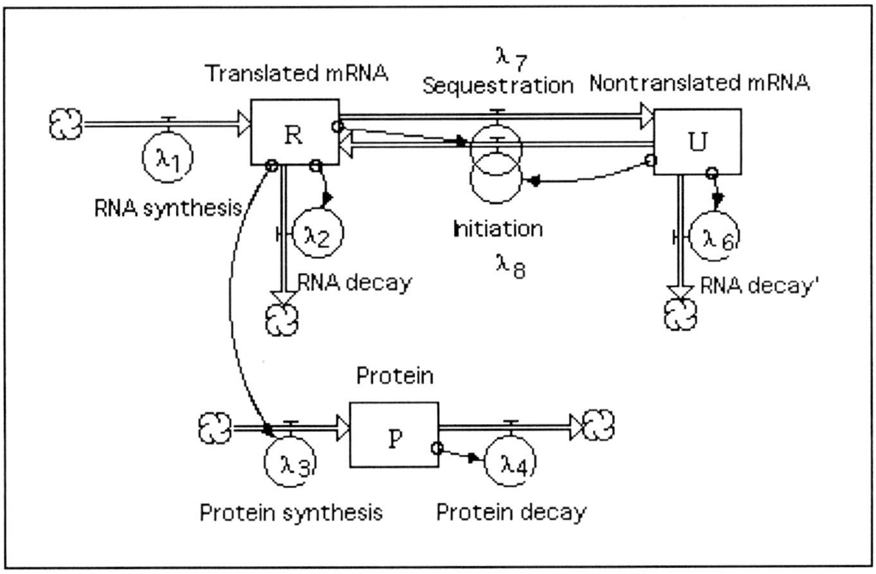

Fig. 4.3 A model including post-transcriptional control is easily constructed by adding an extra compartment for untranslated mRNA and a means of controlling the rate of interconversion. The consequence of sequestering the mRNA is that it need not be synthesized and can begin protein synthesis very rapidly, thus bypassing the delay imposed by accumulation controlled by high RNA stability.

In some cases, a molecule may be depleted from a pool by more than one route. This is true for nuclear mRNA, which may be degraded or exported to the cytoplasm, and for secretory proteins, which may be degraded or secreted. If the outflows are proportional to concentration, then flow through each efflux pathway may be calculated by multiplying the concentration of product times the fractional rate corresponding to each pathway. The effective rate at which the pool is depleted is then equal to the sum of the rate parameters. For example, in Figure 4.1, the effective rate of depletion of the nuclear mRNA equals $\lambda_2+\lambda_3$.

TRANSLATIONAL CONTROL

Numerous examples of translational control mechanisms are now known, for example the regulation of ferritin synthesis by the iron regulatory element,[4,5] and the control of ornithine decarboxylase by the presence of polyamines.[6,7] How can a simulation program be modified to include this process? The new software programs make this very easy to do. For example, the basic diagram may be modified as shown in Figure 4.3, which shows that cytoplasmic mRNA exists in translationally active and inactive forms, designated R and U, respectively. Steady state equations for any diagram in which a component flows reversibly between two compartments can be obtained by setting up simultaneous equations, solving for one component in terms of the others, and substituting to obtain the correct values for use in the computer program. For Figure 4.3, the solution is obtained as follows.

Recall that net flux equals zero in the steady state, and that the quantity in each compartment is a function of the rates of influx and efflux. The rate of protein synthesis will still be a function of the abundance of translated mRNA, as discussed in the prior section, and may be ignored. Note that the amount of mRNA that initiates protein synthesis, R, is formed not only from newly synthesized RNA, but also from the untranslated pool, as represented by the two parameters, λ_1 and λ_8, and is depleted not only by mRNA degradation (λ_2), but also by sequestration into the non-translated pool (λ_7). The mass balance equation for synthesis and degradation of the translated mRNA is: $\lambda_1+\lambda_8U = R\lambda_2+R\lambda_7$. Therefore, the steady state equations can be written as follows:

$$R = \frac{\lambda_1 + \lambda_8 U}{(\lambda_2 + \lambda_7)} \qquad \text{Eq. 4.4}$$

$$U = \frac{\lambda_7 R}{\lambda_6 + \lambda_8} \qquad \text{Eq. 4.5}$$

The initial values for R and U can be solved by substituting the value of U from equation 4.5 into Eq. 4.4. To make this easier, define constants $A = \lambda_1/(\lambda_2 + \lambda_7)$, $B = \lambda_8/(\lambda_2 + \lambda_7)$, and $C = \lambda_7/(\lambda_6+\lambda_8)$. Then

$$U = CR \qquad \text{Eq. 4.6}$$

$$R = A + BU = A + BCR \qquad \text{Eq. 4.7}$$

$$\text{Therefore, } R = A/(1\text{-}BC) \qquad \text{Eq. 4.8}$$

The steady state values for the translated pool can be obtained from Eq. 4.8, and the value for the non-translated pool of mRNA can be calculated from Eq. 4.6. Notice that for any set of kinetic parameters, at steady state the relative amount of RNA in the two pools is $U/R = C$.

If the protein were interconverted between active and inactive forms, the same method would be used to solve for the concentrations of the two forms. After the steady state values have been calculated, the computer program may be used to simulate effects of changing various controls.

In practice, the relative quantities are regulated in a manner that depends on the binding of *cis*-acting factors or other control mechanisms. The software permits this to be modeled as a switch that is either on or off (presence or absence of hemin or free iron in the example of ferritin), or by adding a separate module that represented the concentration of iron and of the iron-dependent regulatory protein. This is easy to do with the simulation package, but the user must be cautious of one foible that is intrinsic to programs based on finite difference equations. If kinetic parameters are expressed as fractional exchange rates per hour, and a control can occur within minutes, then the default calculation interval must be set to a shorter value. As a rule of thumb, the calculation interval should never exceed half of the shortest time constant in the model, or else the system will oscillate erratically. For example, the default calculation interval in the STELLA II program is 0.25 time units, or hours in the present case. If the translational mechanism had a half-time of 5 minutes (0.083 h), then the calculation interval should be changed to a value less than or equal to 0.04 hour.

The importance of translational controls lies in their ability to permit rapid synthesis of new protein without the delays imposed by the other rate parameters.[3] If the half-life of mRNA in the cytoplasm was 2 hours, then first-order kinetics suggests that a transcriptionally active stimulus would require at least 2 hours to achieve half-maximal rates of protein synthesis, and about 10 hours to attain a new rate of synthesis. In contrast, preformed mRNA could become translationally active in a matter of minutes, which is very important in preventing iron-dependent oxidative damage to the cell. The only other alternatives for rapid acquisition of extra capacity would include the use of preformed protein, which is not always feasible due to the many competing needs of the cell, or an increase in the amount of protein synthesized per mRNA molecule. Predictions based on these models will be discussed in the next chapter.

OBTAINING DATA FOR SIMULATION EXPERIMENTS

In practice, acquiring appropriate data is the most difficult aspect of modeling, because it requires making measurements of several distinct processes with consistent units, or measuring enough rates and concentrations to be able to calculate any unknown values. Historically, many studies on proteins preceded work with the corresponding mRNAs, and it is relatively rare to find all the appropriate data in one source. It is also very difficult to measure rates of transcription in living cells because of the very small concentration of specific transcripts in the nuclei, and the difficulty in attaining a high level of radioactive precursor for pulse-chase experiments. Most work employs nuclear run-on transcription, and it is questionable to what extent this represents conditions in vivo. The data used for modeling must at least reflect the fold-increase or decrease in transcription; preferably, the number of transcripts generated per unit of time per cell should be determined. First, it is necessary to measure cell number and the proportion of cells in the population that express the gene of interest. Concentrations of mRNAs can be determined rapidly and accurately by solution hybridization to complementary RNA probes using a version of the RNase protection assay[8,9] or the competitive polymerase chain reaction.[10] Concentrations may then be expressed as molar quantities or as molecules per cell.

The three basic methods of determining half-lives of individual mRNAs and proteins are (1) pulse-chase experiments, (2) analysis of approach to steady state, and (3) rate of degradation after synthesis is terminated with antibiotics.[11] The pulse-chase experiment requires uniform labeling of the precursor pool with a radioisotopic precursor at a high enough specific activity that the product can be detected, and then a chase with a non-radioactive precursor to prevent continued synthesis of radioactive product. The product must then be purified, typically by use of complementary DNA or RNA, or by immunoprecipitation of a protein. This method is affected by reuse of precursor from previously labeled cellular constituents, by possible effects of high concentrations of precursor added during the wash-out period, and by the rate of cell division. Successful use of the technique requires that the gene product of interest attain sufficiently high concentration so that isotope incorporation can be detected. Few cells take up RNA precursors at high rates, and specialized techniques for permeabilization of the plasma membrane may be needed to increase the specific activity of the precursor pool.[12] The approach to steady state is an indirect method of estimating half-lives or rates of decay, but is the least invasive of the three measures.[8,11] It is based on the log-linear relationship between the approach to a new concentration and the rate of elimination of a molecule from the pool of interest (Fig. 2.3).

In kinetic modeling, the quantitative relationships among the rate of transcription and concentrations of various gene products are a matter of scaling. The shape of a theoretical curve based on a specific model of mRNA or protein metabolism is always the same for one set of assumptions about the control mechanisms; the aim of the experimentalist is to define the scale for a system of interest, and this is easiest to do in the steady state. For example, enzyme quantities are usually determined by assaying catalytic activity, and each enzyme possesses a specific amount of catalytic activity per molecule, which cannot be predicted and must be measured. Appropriate kinetic constants for computer-based simulation of enzyme induction may be derived in several ways. As an example, the values used for tyrosine aminotransferase and its mRNA were calculated as follows.[13,14] The unit of enzyme activity is expressed as micromoles of product formed per minute at 37° C, and purified tyrosine aminotransferase has a specific activity of about 600 units per mg of protein. Livers of fasted rats contain about 2-3 units of enzyme activity per gram of tissue, or the equivalent of 4 micrograms of tyrosine aminotransferase. Since the subunit molecular weight is 50,000 grams/mole, the basal concentration equals about 80 pmol of enzyme per gram liver.

The half-life of tyrosine aminotransferase has been estimated to be 1.5 hours in vivo.[12] The rate constant for degradation therefore equals (0.693)/1.5 h, or 0.462 h^{-1}. At steady state, synthesis and degradation are balanced, and the rate of synthesis equals the protein concentration times the rate constant for degradation. Thus, tyrosine aminotransferase is synthesized at a rate of about (80 pmol/g liver)(0.462 h^{-1}), or about 37 pmol/g/h, or 880 pmol/g/d.

The units for protein synthesis may be expressed as moles of protein synthesized per mole of mRNA per day. This value may be calculated from polysome size and the transit time for ribosomes on this mRNA, which is about 2 minutes.[13] Assuming that 12 polysomes bind per mRNA, translation generates about 360 molecules of protein per molecule of mRNA per hour; this is similar to Palmiter's estimate of 550 for ovalbumin in tubular gland cells.[15] Nickol et al[14] estimated the concentration of mRNA for tyrosine aminotransferase to be 0.025 pmol/g of liver, and estimated its half-life to be 1.2 h. The rate of degradation is (0.693)/(1.2 h), or 0.58 h^{-1}. Thus, the rate of mRNA synthesis for tyrosine aminotransferase is at least (0.025 pmol/g liver)(0.58 h^{-1}), or 0.014 pmol of mRNA/g liver/h. This value may be used to estimate a rate of transcription on the assumption that pre-mRNA sequences are conserved. Liver contains about 10^8 hepatocytes per gram, each hepatocyte contains about 150 molecules of this mRNA, and the rate of synthesis per cell is (150 molecules/cell)(0.578 h^{-1}), or about 87 molecules per cell per hour. Because the gene is autosomal, the basal rate of transcription from this gene is approximately 0.7 $minute^{-1}$.

MODELING TRANSCRIPTIONAL CONTROLS

The Activation of Transcription

The original kinetic models of enzyme induction considered only three elements: the concentration of an enzyme, its rate of synthesis, and its rate of elimination. Now that the processes by which transcription is controlled are much better understood, it is legitimate to ask how to generate a realistic model that is accessible to scientists who have a basic grasp of kinetic principles. The aim of this section is to discuss the simplest possible way of accomplishing this task, recognizing that although every gene is controlled by numerous factors, most experimentalists only study one or two at a time.

The idea that enzyme induction could be modeled as "zero order synthesis, first order decay" was known to be an oversimplification when it was first proposed because it omitted details of the processes involved in transcription and translation. Technically, that model is only correct when the rate of synthesis changes instantaneously from one level to another. The model worked relatively well because mRNAs and proteins in mammals are relatively stable once synthesized, with half-lives that usually exceed 1-2 hours and frequently are much longer than this. For instance, the average half-life of mRNA in liver has been reported to range from about 5-20 hours, and the average half-life of hepatic protein is in the range of 3-5 days.[16-18] Because the time course for approach to steady state is limited by the slowest rate parameters in effect, the short periods needed to activate transcription can frequently be ignored. For many mammalian genes, transcription can be activated to its maximum extent within 15-30 minutes. This means that the half-time for transcriptional activation typically represents about 5% of the time needed to achieve a half-maximal increase in mRNA, and less than 1% of the time needed for half-maximal increase in protein. Therefore, the idea that induction can be modeled with the rate of synthesis represented as zero order frequently is a very reasonable assumption. Still, there are numerous examples in the literature in which this model is inadequate because the accumulation curve for mRNA is sigmoidal, and often is associated with a delay that is not predicted from the basic model. Because the events that produce the delay and the sigmoidal shape are often central to understanding the mechanism of control, it is important to discuss how these components can be incorporated into a kinetic model.

Kinetics of Transcriptional Activation

If the binding of proteins to DNA only involved one target DNA sequence and one site on the DNA-binding protein, then the reaction would be expected to follow a binding curve predicted by Michaelis-Menten kinetics, and a plot of the fraction of available binding sites versus concentration of the protein would be expected to be hyperbolic. In practice, it seems unlikely that such simple kinetics will be common for several reasons:

1. Many DNA-binding proteins are dimers or multimers, and the dimerization reaction causes the binding curve to have a sigmoidal shape, whether the curve is expressed in terms of protein concentration or in terms of the time course of the reaction;
2. Many DNA-binding proteins are regulated by ligands or by covalent modification, and the sequence of ligand binding followed by binding of the complex to DNA produces a sigmoidal curve;
3. Multiple binding sites for specific regulatory proteins exist in promoter elements, and this creates the possibility for allosteric effects, cooperativity, and threshold behavior;
4. Cooperative interactions occur not only between DNA-binding proteins and RNA polymerase, but also with other regulatory factors.

Many DNA binding proteins are activated or inactivated either as a result of ligand binding, or because they are controlled directly or indirectly by regulatory enzymes. This means that transcriptional controls maybe simulated by using well established principles for analyzing the interactions of ligands with their receptors, or of substrates with enzymes.[19-23] After a DNA binding protein has been activated, its binding to regulatory elements in DNA also has a kinetic aspect that must be considered.

Rates of Association Between Ligands and their Receptors

The minimum number of processes that are necessary to simulate effects of DNA binding proteins on transcription are two: initial binding of a ligand to its specific receptor, and subsequent binding of the ligand-receptor complex to DNA (or dissociation of the complex from DNA). Examples of this kind of process include the binding of steroidal hormones, second messengers, metabolites, and metal ions to their receptors.

In the case in which a single ligand binds to one site on a receptor molecule, the rate of association and dissociation can be defined from the following equation:[19,20]

$$L + R \underset{k_2}{\overset{k_1}{\rightleftarrows}} LR \xrightarrow{k_3} LR*$$

The rate of association equals $k_1[L][R]$, in which k_1 has units, $M^{-1}time^{-1}$, or liters*mole^{-1} time^{-1}. *Note that if is $k_1$multiplied times ligand concentration, the rate of association ($k_1[L]$)now has units of time^{-1}*. In other words, the time dependency of the response is a function of k_1 and ligand concentration, as has been shown clearly for binding of dexamethasome to the glucocorticoid receptor.[19] The rate of dissociation is first-order; it equals $k_2[LR]$, and k_2 has units of time^{-1}. Although this is the most basic formulation, in some cases the LR complex changes conformation and becomes active for binding, in which case the active complex, LR*, is regarded as a distinct entity with a unique rate constant for formation. For the time being, we will ignore the transition to the active state, and ask only how the concentration of the complex varies with time.

If only the association reaction is considered, then the rates of association and dissociation are equal at equilibrium,

$$k_1[L][R] = k_2[LR] \qquad \text{Eq. 4.9}$$

This is related to the apparent equilibrium association and dissociation constants as follows:

$$K_D = \frac{[L][R]}{[LR]} = \frac{k_2}{k_1} \qquad \text{Eq. 4.10}$$

p. 55, in second line after the first equation: transpose "is k_1" to read "k_1 is."

Therefore, the dissociation constant at equilibrium can be defined either as a ratio of concentrations or as a ratio of dissociation and association rate constants. In the kinetic analysis, the rate constants are more important because they include the dimension of time. Notice that the K_D has units of concentration, and represents the concentration of ligand at which a half-maximal binding response is observed.

How does the LR complex concentration change with time?

$$\frac{d[LR]}{dt} = k_1[L][R]-k_2[LR] \qquad \text{Eq. 4.11}$$

When the concentration of ligand is much greater than R, which is often the case, the rate of approach to equilibrium equals

$$\ln\frac{[LR]eq}{([LR]eq^{-}[LR]_t)} = (k_1[L] + k_2)t \qquad \text{Eq. 4.12}$$

In other words, the rate at which the LR complex forms depends on both the forward and reverse kinetic constants and on the concentration of ligand. A plot of the function on the left side of this equation over time gives a hyperbolic curve when the ligand is present at high concentration, and the logarithmic function is a straight line.

An alternative way of expressing this information in terms of ligand concentration is to consider the binding at any concentration of ligand as a fraction of the total possible binding in the tissue. This can be expressed similarly to the Michaelis-Menten expression as follows:

$$[LR] = \frac{[L][R]total}{K_D+[L]} = \frac{[L][LR]max}{K_D+[L]} \qquad \text{Eq. 4.14}$$

because the maximum amount of complex is equal to the total receptor present in the system. As with the Michaelis Menten equation, this represents a rectangular hyperbola that is half maximal when the ligand concentration equals the value of the K_D. This is actually a single solution to a more general equation in which there is cooperativity among binding sites for multimeric units. In this case, the Hill equation represents a more general solution.[22] Basically, it represents the fraction of ligand that is bound in an active complex at any concentration, multiple times the maximal possible complex:

$$[LR] = \frac{[L]^n[LR]max}{[L_{0.5}]^n + [L]^n} \qquad \text{Eq. 4.15}$$

where the concentration of ligand giving half-maximal response is defined as $L_{0.5}$, and equals the K_D. The exponent n can have any positive value that does not exceed the number of binding sites; it need not be an integer. The fractional amount of LR present at any time is equal to:

$$\frac{[LR]}{([LR]max^{-}[LR]} = \frac{[L]^n}{L_{0.5}{}^n} \qquad \text{Eq. 4.16}$$

Strickland and Loeb[23] have employed a similar kind of analysis to link the generation of second messengers to receptor saturation and subsequent biological response.

LAGS AND THE IDEA OF RELAXATION TIME

Just as musical compositions have time signatures, every complex biological process has duration and requires predictable periods to approach a new equilibrium. The aggregate rate for the process depends on the sum of the rates of the constituent processes, and some contribute more to the overall rate than do others. Experimentally, the process of identifying the contributions of each sub-process yields important information about mechanisms. This idea has received theoretical treatment in the concept of "relaxation times," by which the biochemist is able to identify all those processes which contribute to the time needed for a reaction to approach its equilibrium.[2] In this technique, a change is introduced, and the time required for the reaction to seek a new equilibrium is determined. The system proceeds to this equilibrium via a series of relaxation times, τ (= the reciprocal of the rate constant[s]). Effectively, all the identifiable rates are added together to achieve a global value that depends on the sum of its parts. And because each step in gene expression has a magnitude, the initial value may be amplified or attenuated, thus introducing the idea of scaling into the picture. The hierarchical organization of the human and animal body makes this last idea critical, for the processes initiated by one or two genes per cell must be magnified to change a property, or phenotype, of the whole organism.

Often, gene expression is induced by addition of a high concentration of ligand such as a steroidal hormone or a cyclic AMP analog, and the contribution of the ligand concentration to time course is ignored. Even under these circumstances, it is common to find that the rate of transcription from different genes, as well as rates of mRNA and protein accumulation, differ. What mechanisms can account for these differences? Almagor and Paigen[1] have modeled induced gene expression as a function of binding of *trans*-acting factors to promoter elements, and concluded that the minimum expression needed to simulate the time course includes three separate functions, namely, the rate of association and dissociation of the binding protein from the DNA, the half-life of the mRNA, and the half-life of the protein. In addition, their model included the supposition that several binding sites for one factor may exist on a single gene. The model assumed that the activated receptor complex bound to a DNA element designated s, and converted it to a DNA-receptor complex designated s*.

$$s \underset{k_2}{\overset{k_1}{\rightleftarrows}} s^*$$

where both rate constants have units of time^{-1}. When there are n different binding sites for the receptor, the fraction of the genes with a given number of bound receptors after addition of the inducing agent is expressed as follows:

$$X_t = \frac{k_1}{(k_1 + k_2)} \left(1 - e^{-(k_1 + k_2)t}\right) \qquad \text{Eq. 4.17}$$

The sum of the two rate constants, k1 + k2, was defined as the relaxation time for binding, and was designated k^0. In other words, the rate at which a specific fraction of DNA binding elements was occupied was determined by the sum of the two rate constants. The average occupancy of the sites in a population of cells is then equal to the number of sites available on each promoter (N) times X_t, and the rate at which this occupancy is attained is a simple exponential function.

To duplicate the effects of this mechanism on gene expression, it is necessary to add to this the equations for accumulation of mRNA and protein, so that the result has the following form:

$$P_t = P_{ss} - Ae^{-k_o t} - Be^{-k_r t} - Ce^{-k_p t} \qquad \text{Eq. 4.18}$$

where A, B, and C are constants related to the kinetic constants, and P_{ss} equals the mRNA concentration at steady state multiplied times the rate of protein synthesis per unit of mRNA, and divided by the rate of protein degradation. Effectively, this is the sum of three exponential curves, and has a sigmoidal shape. The main conclusion based on the consideration of ligand-dependent activation of a receptor that binds to DNA, and the subsequent binding of the complex to DNA, is that effective simulation of gene activation must include at least one saturation curve, and this introduces a time dimension into the activation process that in some cases amounts to several hours. The temporal response of the entire process includes the activation phase and period needed for mRNA and protein to accumulate. Other mechanisms that can also produce delays include cooperative binding at multiple sites, thresholds for activation of transcription, and physical changes

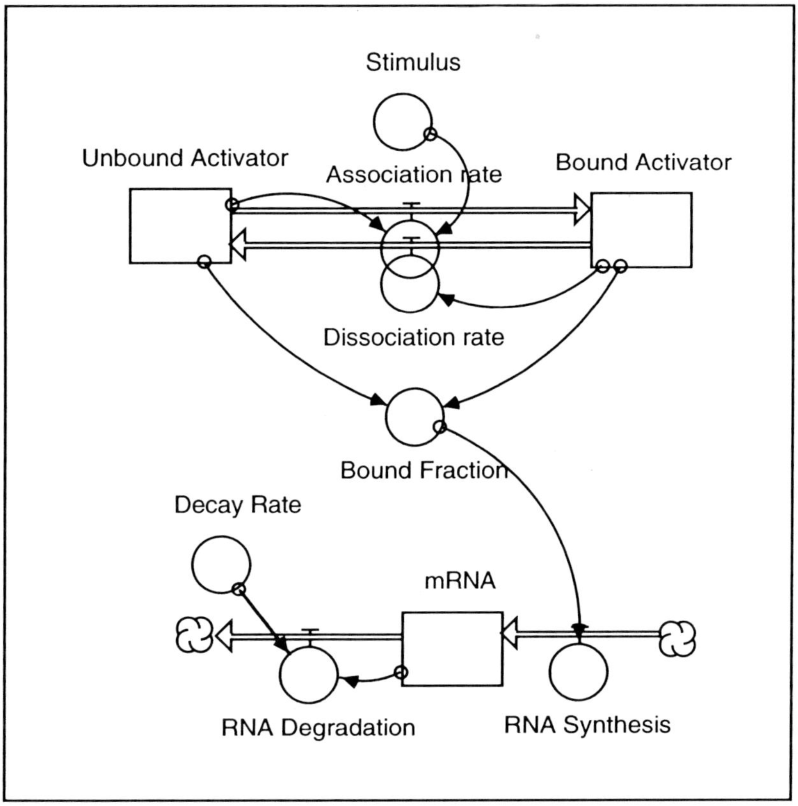

Fig. 4.4. The activation of DNA-binding proteins also adds a delay to initiation of transcription. The rate of activation depends not only on the rate at which the active complex binds DNA, but also on the rate at which the DNA-binding factor becomes competent to bind.

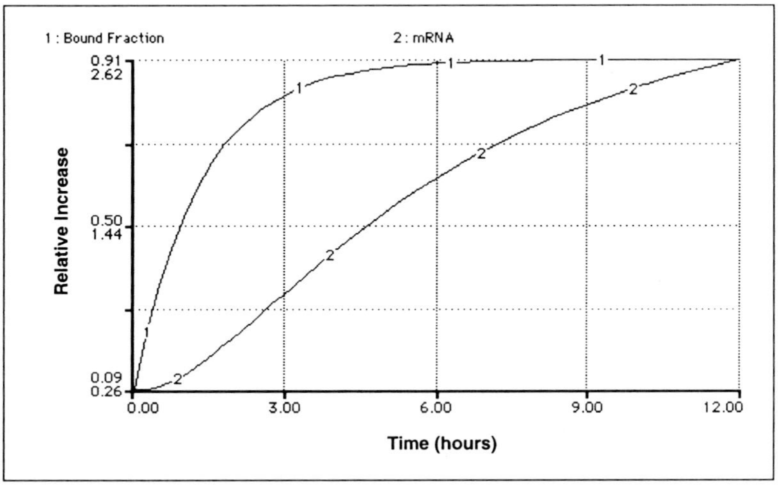

Fig. 4.5. Simulation of transcriptional activation by a trans-acting *factor indicates the delay imposed by the time required to achieve a maximal bound fraction. The consequence is to cause a delay in mRNA induction, which produces a sigmoidal accumulation curve for the mRNA.*

in the initiation complex.[1,25-28]

The description of how ligands and receptors interact provides a means of using a computer program such as STELLA II to link the binding of a *trans*-acting factor to a gene, altered transcription, and accumulation of mRNA or protein products as a function of time and ligand concentration. In the example diagrammed in Figure 4.4, the rate of mRNA synthesis was assumed to be a function of the fraction of a regulatory molecule that was bound to the promoter. As shown in Figure 4.5, this introduces a delay compared to the single compartmental model, so that the predicted mRNA response curve is sigmoidal even if nuclear processing is ignored, as suggested by Almagor and Paigen.[1] The length of the delay depends on the rate at which the *trans*-acting factor was activated, and would be negligible if the activation was very rapid compared to the kinetics of mRNA accumulation. Any of these equations may used in modeling programs to achieve outcomes that simulate experimental data.[24] In the STELLA II program, this is done by typing the equation into a dialog box associated with a control element such as the icon for RNA synthesis in Figure 4.4.

THE LAW OF DIMINISHING RETURNS

Finally, if one is considering a protein that is an enzyme, it is necessary to ask what effect a change in enzyme concentration has on flux of a substrate through a pathway. The change in flux as a function of enzyme activity cannot be predicted in advance. It depends on characteristics of the enzymes in question, the quantities of each in the cell under initial conditions, and their kinetic parameters such as affinities, half-maximal velocities, and influence of products and allosteric modulators.[29,30] In well-studied systems, the measured responses can be plotted as hyperbolic curves, with small changes from the initial values producing relatively large effects on flux. However, as enzyme concentrations increase, the effect on flux becomes smaller and smaller, approaching an asymptote of zero.

Hepatic metabolism of amino acids is affected by the activity of enzymes that are induced by dietary protein and by hormones such as cortisol and glucagon. In situations in which diets contain moderate levels of protein (10-20% by weight), the enzyme activities are low, but the activities increase by five to tenfold as dietary protein is increased, and by similar values when animals are treated with glucocorticoid hormones that mimic the action of cortisol or corticosterone. In a study of aromatic amino acid metabolism in brain and liver, Salter et al[31] found that metabolism of tryptophan and tyrosine was affected both by the rates of transport into cells and by the concentrations of specific enzymes in the metabolic pathways by which these amino acids were metabolized. They measured the flux as a function of the activities of tryptophan oxygenase and tyrosine aminotransferase. The data for the change in flux as a function of enzyme quantity fit hyperbolic curves of the following form:

$$J = \frac{J\text{max } V}{(V + K)} \qquad \text{Eq. 4.19}$$

where J is observed flux under experimental conditions, J_{max} is predicted flux at infinite enzyme velocity, V is assayable maximum activity for the enzyme in question when assayed in vitro, and K is a constant defined as the value of V when $J = J_{max}/2$.

If one intends to model the effects of altered transcription on the ability to metabolize an amino acid, the consequences of this theory are important. Firstly, it is clear that a specific increase in transcription may eventually produce an equivalent increase in enzyme concentration, and therefore an increase in the assayable enzyme activity. This does not mean that flux will change in proportion. Instead, the Law of Diminishing Returns holds, because the more the enzyme activity increases, the less it controls the flux through a particular metabolic pathway. Eventually, the rate of metabolism will be limited by substrate availability, by the rate of another enzyme in the pathway, or by accumulation of product. In fact, this has a very practical implication for the mammalian cell. For enzymes that control the rate of reactions but that are present at significant amounts in the "basal" condition, small increases in concentration will effectively increase the rate of metabolism. However, an increase in concentration beyond a certain level is ineffective. In the data obtained by Salter et al,[31] the limiting enzymes controlled about 70% of flux though the pathway in the basal state, but only about 10-20% of flux in the induced state. If the point of regulation is to achieve altered rates of metabolism, plainly it is not effective in cases such as this to increase transcription more than tenfold. This is important for cellular economy, because cell size and resources for protein synthesis are limited, and there are competing demands for enzymes in other pathways.

References

1. Almagor, H, Paigen K. The chemical kinetics of induced gene expression. Activation of transcription by noncooperative binding of multiple regulatory molecules. Biochem 1988; 27: 2094-2102.
2. Fersht A. Enzyme structure and mechanism New York: WH Freeman, 1985.
3. Hargrove JL. Microcomputer assisted kinetic modeling of mammalian gene expression. FASEB J 1993; 7: 1163-1170.
4. Munro HN. Iron regulation of ferritin gene expression J Cell Biochem 1990: 44:107-115.
5. Hentze MW, Rouault TR, Harford JB, Klausner RD. Oxidation-reduction and the molecular mechanism of a regulatory RNA-protein interaction. Science 1989: 244:357-359.
6. Persson L Hom I, Heby O. Regulation of ornithine decarboxylase mRNA translation by polyamines. J Biol Chem 1988; 263:3528-3533.
7. Manzella JM, Blackshear PJ. Regulation of rat ornithine decarboxylase mRNA translation by its 5'-untranslated region. J Biol Chem 1990; 265:11817-11822.
8. Durnam DM, Palmiter RD. A practical approach for quantitating specific mRNAs by solution hybridization. Anal Biochem 1983; 131:385-393.

9. LeCras TD, Gherardi E, Bowyer DE. A sensitive RNase protection assay for the quantitation of the mRNAs for the LDL receptor and HMG CoA reductase in human total RNA. Atherosclerosis 1991; 90:81-90.
10. Gilliland G, Perrin S, Blanchard K, Bunn HF. Analysis of cytokine mRNA and DNA: detection and quantitation by competitive polymerase chain reaction. Proc Natl Acad Sci USA 1990; 87: 2725-2729.
11. Rodgers JR, Johnson ML, Rosen JM. Measurement of mRNA half-live as a function of hormonal treatment. Meth Enzymol 1985; 109: 572-580
12. Ucker DS, Yamamoto K. Early events in the stimulation of mammary tumor virus RNA synthesis by glucocorticoids. Novel assays for transcription rates. J Biol Chem 1984; 259: 7416-7420.
13. Kenney FT, Lee KL. Turnover of gene products in the control of gene expression. BioScience 1982; 32: 181-184.
14. Nickol JM, Lee, K-L and Kenney FT. Changes in hepatic levels of tyrosine aminotransferase mRNA during induction by hydrocortisone. J Biol Chem 1978; 253: 4009-4015.
15. Palmiter RD. Quantitation of parameters that determine the rate of ovalbumin synthesis. Cell ,1975; 4: 189-197.
16. Moore RE, Goldsworthy TL, Pitot HC. Turnover of 3'-polyadenylate-containing RNA in livers from aged, partially hepatectomized, neonatal and Morris 5123C hepatoma-bearing rats. Cancer Res 1980; 40: 1449-1457.
17. Schimke RT. Control of enzyme levels in mammalian tissues. Adv Enzymol 1973; 37: 135-187.
18. Hargrove JL, Hulsey MG, Beale EG. The kinetics of mammalian gene expression. BioEssays 1991; 13: 667-674.
19. Baxter JD, Tomkins GM. Specific cytoplasmic glucocorticoid hormone receptors in hepatoma tissue culture cells. Proc Nat Acad Sci USA 1971; 68: 932-937.
20. Taylor P, Insel PA. Molecular basis of pharmacologic selectivity In: Principles of Drug Action The Basis of Pharmacology ed. Pratt WB, Taylor, P. New York: Churchill Livingstone, 1990.
21. Levitzki, Alexander. In: Receptors: A Quantitative Approach, Menlo Park CA: Benjamin/Cummings Publishing Co, 1984.
22. Koshland DE Jr. Switches, thresholds and ultrasensitivity. TIBS 1987; 12;225-229.
23. Strickland S, Loeb JN. Obligatory separation of hormone binding and hormone action. Proc Nat Acad Sci USA 1981; 78: 1366-1370.
24. Hargrove JL, Hulsey MG, Summers AO. From genotype to phenotype: kinetic modeling of gene expression with STELLA II. BioTechniques 1993; 15: 1096-1101.
25. Yamamoto KR, Alberts B. The interaction of estradiol receptor protein with the genome: an argument for the existence of undetected specific sites. Cell 1975; 4: 301-310.
26. Mulvihill ER, Palmiter RD. Relationship of nuclear estrogen receptor levels to induction of ovalbumin and conalbumin mRNA in chick oviduct. J Biol Chem 1977; 252: 2060-2068.
27. Palmiter RD, Moore PB, Mulvihill ER and Emtage S. A significant lag in the induction of ovalbumin messenger RNA by steroid hormones: a receptor translocation hypothesis. Cell 1976:; 8: 557-572.

28. Yagil G. Enzyme Induction. In: Mathematical Models in Molecular and Cellular Biology. Segel LA, ed. Cambridge: University Press, 1980.
29. Westerhoff HV, Groen AK, Wanders RJA. Modern theories of metabolic control and their application. Bioscience Reports 1984; 4: 1-22.
30. Liao JC, Delgado J. Advances in metabolic control analysis. Biotechnol Prog 1993: 9: 221-233.
31. Salter M, Knowles RG, Pogson CI. Quantification of the importance of individual steps in the control of aromatic amino acid metabolism. Biochem J 1986; 234: 635-647.

CHAPTER 5

The Interrelationship of Messenger RNA and Protein Metabolism

If the nucleic acid message may be compared with the text of a book, the regulatory network determines which pages are to be read at any given time. In the expression of the nucleic message, as well as in its reproduction, adaptation results from an elective rather than an instructive effect of the environment.

—François Jacob

The simple mathematics described in this volume has value because it enables one to make a number of testable predictions about the relationship between mRNA metabolism and protein metabolism. The predictions must be explained from the perspective that nothing in biology came about totally by chance, but resulted from organic evolution and natural selection. This is just as true for processes of mRNA and protein degradation as for processes of biosynthesis, because these events require highly regulated and coordinated activity of nucleases and proteases. Just as the biosynthesis of mRNA is normally tightly coupled to biosynthesis of protein, so must degradation of these molecules be coordinated to terminate the effects of any inducing agents. If the timing of a response to an environmental challenge is critical, then the system must acquire rapid and specific mechanisms for controlled degradation of macromolecules, as well as systems for regulation by covalent modification. It is now evident that the means to accomplish this depends on molecular mechanisms of incredible sophistication.

THE NECESSARY BALANCE OF SYNTHESIS AND DEGRADATION

In adult mammals, the rates of total mRNA synthesis and protein synthesis are precisely balanced with the rates of mRNA degradation and protein degradation. For instance, the body weight of adult humans typically varies by no more than 1 pound per year between the ages of 20-50. Although this observation may seem commonplace, it is very difficult to achieve because the processes involved in the biosynthesis of macromolecules are chemically and mechanistically distinct from the processes of degradation. In other words, the body weight of a healthy, adult human is very nearly in a steady state, and extensive physiological mechanisms have evolved to support this balance. Although considerations of mass balance are very simple, they disguise an underlying complexity. The cells that make up each organ individually synthesize from 3,000 to 12,000 different gene products, and the processes involved in mRNA synthesis and protein synthesis are elegantly complex. Knowing this, is it likely that the degradative processes are any less elegant? The balance equations discussed in Chapter 4 indicate that the total amount of mRNA transcription, total production of mature, cytoplasmic mRNA, and the synthesis of all the proteins needed to enable each organ to carry out its vast repertoire of functions are precisely balanced by the total amount of degradation. On a typical day, the balance is exact to within a fraction of a percent, and if the individual is healthy and well nourished, the balance continues for years. If this were not so, the individual would continue to grow, often with deleterious results if nutritional requirements exceeded availability. At the cellular level, the choice for growth is between hypertrophy and cell division, and each process eventually becomes disadvantageous to a mammal, if not catastrophic.

What is the simplest invention that would effectively bring the system into balance? The answer is to achieve a rate of product degradation that is proportional to the amount of product present—in other words, first order elimination! This automatically puts limits on growth, and can be achieved in one of three ways: either the cell must divide, the mRNA and proteins must be secreted, or the gene products must be eliminated by chemical degradation and recycling. Bacteria are unicellular organisms, and the content of most proteins is balanced by a combination of rapid mRNA decay and frequent cell division. Adult mammals, which do not have the luxury of unlimited growth, utilize rates of mRNA and protein degradation that effectively balance synthesis. From this perspective, proportional rates of product elimination had to be invented, and the task may have been just as difficult as developing precise mechanisms of cell cycle control. Despite the well-known difficulty in obtaining stable preparations of most mRNAs and some proteins, the molecules are extremely stable in the absence of degradative enzymes when protected from oxidative destruction. These considerations lead to a series of predictions about mRNA and protein metabolism that are based on the essentials of cellular and organismal homeostasis, plus chemical balance and optimal cell size.

TRANSCRIPTIONAL CONTROLS ARE MOST EFFICIENT

Mathematically, controls over synthesis or degradation can produce equivalent changes in protein concentration, but transcriptional controls are most efficient due to the small scale involved. Each autosomal gene is present at a very low concentration (two copies per cell), and the use of sequence-specific DNA binding proteins permits one variety of protein to control many genes. The amplification involved in gene expression is enormous, because genes are present at a concentration of not over 10^{-12} moles per kg of tissue. This estimate was obtained by dividing the number of hepatocytes in liver (about 10^{11} hepatocytes per kg) by Avogadro's number. Whereas each of the two alleles at a genetic locus can be regulated by a single molecule of a DNA binding protein, the corresponding mRNA may be represented by hundreds or thousands of copies, and therefore has a concentration in the range of 10^{-10} moles/kg. A factor that bound mRNA stoichiometrically would need to be present at a much higher concentration than a DNA-binding protein to achieve the same effect on protein synthesis. Because of the amplification that takes place during protein synthesis, it is plainly not feasible to control protein turnover by stoichiometric binding to a second protein except in special cases. Liver contains about 0.16 g of protein per gram tissue; assuming an average molecular weight of 40,000 grams per mole, the concentration of an average protein equals 4×10^{-7} moles per kg. The relative abundance of proteins compared to genes in liver is therefore about 10^{6} moles of protein per mole of gene. Enzymes control the concentration of metabolites which may range from about 10^{-6} to 10^{-3} moles/kg. The relative excess of metabolites relative to enzymes makes it feasible to control proteins by ligand binding or covalent modification. Both methods are used to regulate cholesterol synthesis by hydroxymethylglutaryl (HMG) CoA reductase, which is unstable in the presence of sterols, and is inactivated by an AMP-dependent protein kinase.[1,2] From the perspective of scale, an enzyme is a gene's means of controlling utilization of metabolites that are present in vastly greater concentration, as appropriate to energy metabolism.

There is a drawback to controls that alter transcription alone. Because mammalian cells do not divide frequently, and because most mRNAs and proteins are relatively stable, the pre-formed molecules persist in tissues for several hours or days after synthesis is terminated. When it is essential that timing be precisely controlled, covalent modification or targeted degradation must be employed to remove the pre-existing products. Bacteria have other options because rapid cell division dilutes pre-existing proteins, and it is evident that mammals employ many more kinds of post-transcriptional regulation than do bacteria. An excellent example is cyclic AMP-dependent protein kinase, which controls glucose metabolism in mammals by regulating glycogen synthase and glycogen phosphorylase.[3] Instead of using a protein kinase for this regulation, most microbes employ a cyclic AMP-binding protein called catabolite activator protein that controls glucose utilization through effects on transcription.[4] Kinetic considerations suggest that mammals invented protein kinases and phosphatases in order to solve the problem of slow adaptation that is imposed by a long life span. Glycogen must be mobilized in seconds during an emergency, and this cannot be achieved by synthesizing an enzyme with a half-life of many hours. A fundamental principle of systems theory is that delays between a signal and a response beget instability.

Time Course of Enzyme Induction Is Independent of Stimulus Magnitude

If the rate of mRNA or protein degradation is truly proportional to concentration, or first-order, then the time required to achieve a new steady state is always the same no matter what the absolute increase in synthesis. Within limits, it takes no longer to increase a protein's concentration by tenfold than by twofold (unless some other resource became limiting). Figure 5.1 demonstrates the theoretical curves for a protein with a half-life of 4 hours that is encoded by a mRNA with a 2-hour half-life. Observe that the half-maximal increase in protein concentration always occurs at the same time despite differences in the absolute increase in synthesis. This is easiest to see in the inset to Figure 5.1, in which the fractional increase was plotted The three curves for twofold, fivefold, and tenfold increases in synthesis correspond exactly when plotted as relative increases. In principle, the protein or mRNA's half-life is also the major determinant of the time course for concentration to decrease when the rate of synthesis diminishes. The idea that half-lives govern the rate of change is thought to be valid except for gene products that accumulate to levels at which synthesis

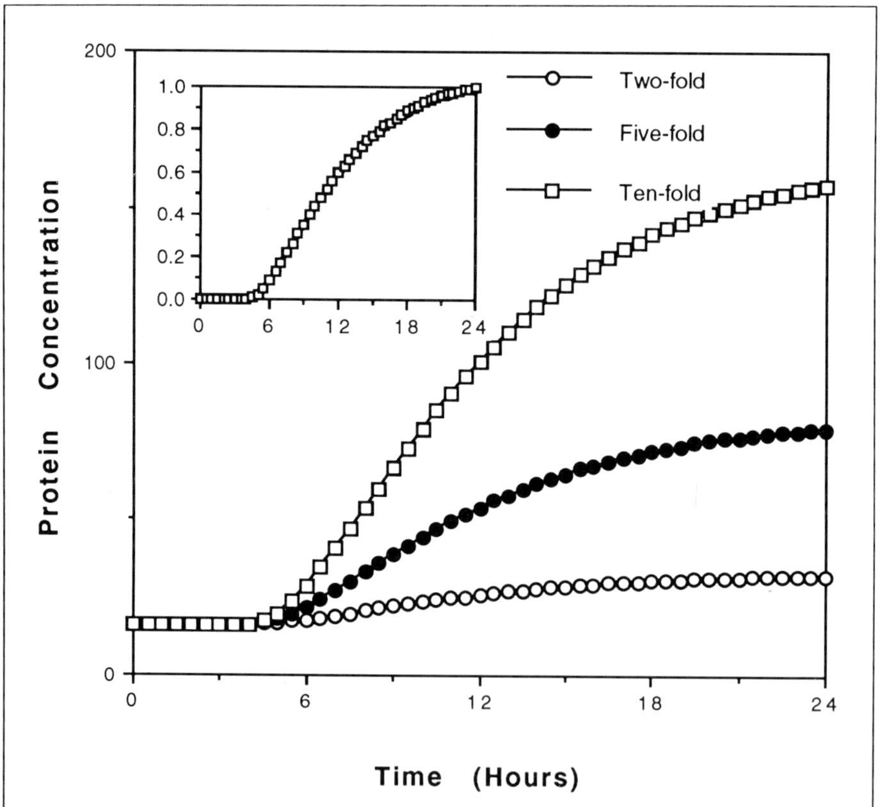

Fig. 5.1. The period required for induction of a mRNA or protein is predicted to be proportional to its half-life, irrespective of the magnitude of the transcriptional stimulus, as shown for three levels of increase. The inset replots the data as relative increases instead of absolute change, in which case the curves are superimposable.

becomes limiting, or decay takes place at a greater rate, and if this were true, a different model would need to be employed. For instance, the logistic model used in population biology limits growth above certain levels. The limitation on cell size and protein content suggest that the logistical model may be appropriate for instances such as over expression of a gene product in transfected cells.

RAPIDLY INDUCIBLE PROTEINS ARE ENCODED BY LABILE mRNAS

Because the approach to steady state depends on a gene product's half-life, a delay will always occur between the time a mRNA attains its half-maximal response and the time when the encoded protein attains its half-maximum value. The mRNA concentration is usually a good index of protein synthesis, but it is akin to a faucet that is filling the protein pool, and not the amount of material already in the pool. Plainly, much more time is required to fill a large pool (stable protein) than a small one (unstable protein). Rapid degradation of mRNAs and proteins is an adaptive feature that permits rapid regulation of concentration, but to say this begs the question of how relative stability evolved. To the degree that stability depends on physical structures and enzyme-based mechanisms such as ubiquitin-dependent protein degradation, then these features must have evolved by natural selection to facilitate interaction of specific gene products with degradative systems. Figure 5.2 gives a very basic example of a mechanism that trades off between product concentration and rapid regulation. Figure 5.2A shows the very short predicted delay between an increase in mRNA and increase in the concentration of an unstable protein. Compared to this, the stable protein in Figure 5.2B is greatly delayed and only attains its final concentration much later due to its long half-life and high concentration in the fully induced state. This is the theoretical result. The actual half-lives of several mRNAs and their encoded proteins are plotted in Figure 5.3. In this figure, relative stability is related to the time required to complete cell division, about 20 hours. Two features are evident. Firstly, the average mRNA is less stable than the protein it encodes; secondly, most proteins in this group with half-lives less than 20 hours are encoded by mRNAs with half-lives under 10 hours, thereby reducing the delay between increase in mRNA concentration and increase in protein concentration.

The magnitude of these delays is important, as is clear if the periods are compared to the time needed to complete each step involved in generating mRNAs and proteins. The periods required for the mammalian enzyme, phosphoenolpyruvate carboxykinase (PEPCK), will be used as an example. Transcription from the gene for PEPCK attains a maximum within 30 minutes after treatment of hepatoma cells with cyclic AMP analogues, and declines at a similar rate after addition of insulin.[5-7] The rate of transcription of this gene by RNA polymerase II has been estimated to be about 1,000 nt/min in the presence of insulin and 2,500 nt/min in the presence of cyclic AMP.[7] A period of about 2-5 minutes is required to complete transcription for this gene, which is 6 kb long, and generates a mature transcript of about 2.6 kb.

Transcription of other class II genes has been estimated to proceed at about 3,000 nt/min;[8] the time required to complete a round of transcription for most genes in vertebrates is less than 10 minutes.

Processing and transport of mRNA from the nucleus delay the accumulation of product in the cytoplasm. In the case of the mRNA for PEPCK, the first mature mRNA appeared within 10-20 minutes after addition of cyclic AMP, indicating that processing of the primary transcript required no more than 10 minutes.[6] The concentration of this mRNA was maximal in nuclei by 45 minutes after treatment, but it continued to increase in the cytoplasm for about 3 hours. The period required to translate mRNAs is similar to the time needed

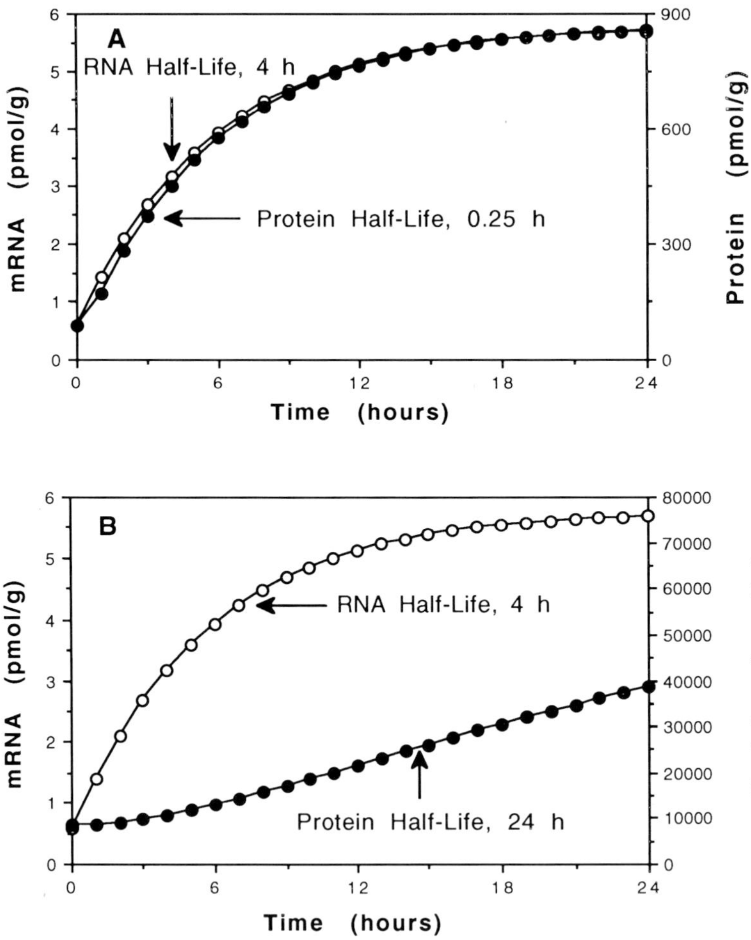

Fig. 5.2. The interval required for an increased level of mRNA to cause a proportional increase in concentration of the encoded protein is predicted to be related to the protein's half-life. (A) For a protein with a short half-life, induction is closely coordinated; (B) For a protein with a long half-life, a long delay is predicted because the respective half-lives govern the period for induction.

to complete transcription; transit time of ribosomes on typical mRNAs is about 2-3 minutes.[9] Even so, a minimum of 12 hours was needed to attain a new, steady-state concentration of PEP carboxykinase.[10] Thus, nuclear processes account for less than 10% of the time needed for full expression of this gene.

The idea that decay constants control the time course of induction to new steady-state levels applies independently to mRNAs and proteins because these molecules represent separate pools. Because most proteins are more stable than the corresponding mRNAs, the time required for proteins to increase to new levels will be delayed in comparison to their mRNAs by a value related to the half-life of the protein.[11-15] This is confirmed by experimental results. The mRNA for tyrosine aminotransferase begins to increase within a few minutes after treatment of hepatoma cells with dexamethasone phosphate (a synthetic, glucocorticoid hormone), and within two hours this gives rise to an increased concentration of the enzyme.[16,17] In contrast, the mRNA for glucose 6-phosphate dehydrogenase begins to increase 5 hours after feeding starved rats a high-carbohydrate diet, doubles by 10 hours, and reaches its final value after about 24 hours.[18,19] The first detectable increase in protein *synthesis* takes place at about 10 hours, and a new steady state protein *concentration* is achieved only after 2-3 days.[18-20] The reason for the pronounced delay prior to induction of the mRNA is not clear, but transcription begins to increase only after 5 hours and attains a peak value at 10 hours. Induction of malic enzyme is similarly delayed relative to the mRNA after treatment of rats with

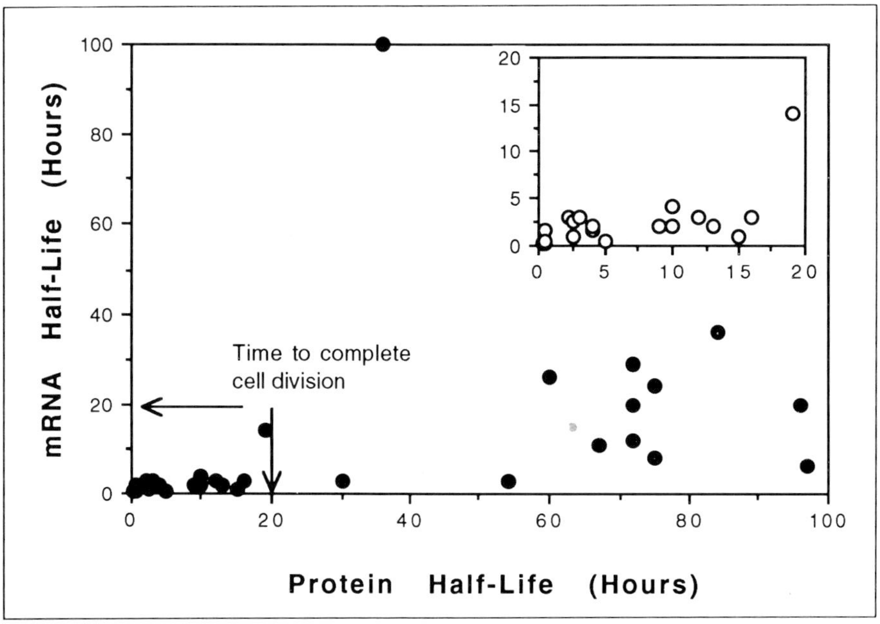

Fig. 5.3. Proteins with short half-lives are usually encoded by mRNAs with short half-lives. The inset shows data for proteins with half-lives of less than 20 hours, which typically are encoded by mRNAs with half-lives of 1-5 hours.

thyroid hormone or feeding a high carbohydrate diet.[21] Due to the wide variation in stability, the predicted delay ranges from a few minutes to more than 60 hours for different pairs of mRNA and protein. Note that the delay applies to the change in protein concentration, not to the rate of synthesis, which is often an excellent index of mRNA concentration.[16,21,22]

Periods required to activate transcription from many vertebrate genes are similar to those reported for PEPCK, but the delay prior to full gene expression may be extreme. An example is the gene for alcohol dehydrogenase in mouse kidney; androgen elicited an increased rate of transcription that reached a peak value only 2-3 days after pellets of testosterone were implanted.[23] In this example, mRNA and enzyme concentrations peaked after 3-4 days and 6-8 days, respectively. The major determinants of these protracted induction periods are the half-lives of the respective mRNAs and proteins; there is excellent agreement between the theory and experimental results in many cases.

DO EXONS COORDINATE mRNA AND PROTEIN STABILITY?

The selective forces that generated the present systems probably included the need to obtain a timely adaptive response for important regulatory proteins, and the need to balance the total rates of synthesis and degradation to achieve optimal cell size and function. The reason for this is that the same kinetic principle applies to inducible proteins and mRNAs: the time required to achieve a new steady state concentration is related to the molecule's chemical half-life. Part of the selective pressure that would produce this result is that a signal is not effective unless it can be turned off. Therefore, when transcription from a gene decreases in a non-dividing cell, the corresponding protein's concentration will decrease according to its half-life. Much is being learned about the structural features that permit this regulation, and it is possible to suggest ideas about this coordination between metabolism of specific mRNAs and their encoded proteins.

If one accepts the idea that the chemical stability of RNAs and proteins depends on specific structural features, the question can be rephrased to ask how the structural features of chemically different macromolecules might be coordinated to achieve rapid enzyme adaptation. One answer to this is to ensure that the destabilizing features are encoded on the same exons. If a single exon could confer a short or long half-life on an mRNA and the corresponding protein, then coordination would be ensured.

Many systems exist in which a mRNA and the encoded protein are both unstable, notably among various protooncogenes and highly regulated enzymes such as ornithine decarboxylase and hepatic enzymes that respond to diet. Figure 5.4 compares the structure of the cellular protooncogene, c-*fos*, with that of the viral form, v-*fos*.[24,25] The last exon in the c-*fos* gene includes coding segments for the carboxylic end of the molecule, as well as the 3', nontranslated portion of the mRNA. The end of the protein contains several structural elements that are commonly found in unstable proteins; these regions contain high contents of proline, glutamic acid, serine, and threonine, and have been termed PEST elements.[26] The same exon in the mRNA includes several destabilizing elements that contain UUAUU, represented in the figure as U_nA.[27,28] These elements are common in cytokine mRNAs, and the mRNA becomes much more stable when these units are removed by deletion. Now compare this with the viral oncogene, which has altered the PEST elements by mutation, and has deleted the portion of the exon that contains the U_nA features. The kinetic model predicts that these changes will cause a given rate of transcription to produce a much higher concentration of mRNA and protein, and will produce a switch that cannot be turned off.

As a second example, consider the transaminase superfamily of pyridoxal-5'-phosphate dependent enzymes. Aspartate aminotransferase and alanine aminotransferase are both very stable gene products which are encoded by stable mRNAs.[29-32] This contrasts with the tyrosine aminotransferase gene, which produces an unstable mRNA and a labile protein that is rapidly inducible by glucocorticoid hormones.[31] Figure 5.5 compares the structures of aspartate aminotransferase and tyrosine aminotransferase; the labile enzyme is similar to the stable enzyme, except that it includes segments at the N- and C-ends which are not represented in aspartate aminotransferase.[33-35] Both of these segments in tyrosine aminotransferase can be phosphorylated, and a basic segment near the N-terminus is readily cleaved by several proteases.[34] Recently, Stellwagen[36] has demonstrated that tyrosine aminotransferase is stabilized by small deletions in either of these two sites that do not affect the enzyme's catalytic activity. It is known that the two segments are encoded by the second and last exons in this gene. The last

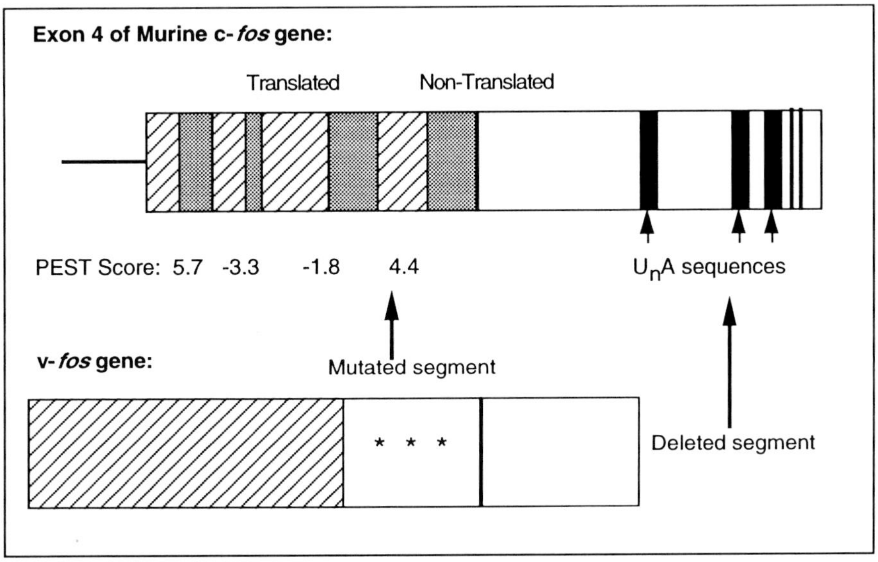

Fig. 5.4. Exon 4 of the murine c-fos gene encodes segments of the protein that are thought to be destabilizing (PEST elements), and also the 3'-untranslated region with its destabilizing U_nA segments. In contrast, the viral gene, v-fos, has mutated or deleted these segments, which cause it to be stable and expressed at high levels.

exon also contains U_nA segments, however the structural basis for the mRNA's instability has not yet been tested. It is interesting that the unstable tyrosine aminotransferase appears to have acquired a short chemical half-life by the addition of specific structural elements that are separate from the features that permit it to transaminate tyrosine, but that enable rapid regulation in response to hormones or diet.

CAN THE KINETIC RULE BE CIRCUMVENTED?

If it is imperative that the concentration of a stable gene product be changed quickly, at least one means is available. The decay constant will always govern the period required to shift between two equilibrium values, but the concentration can be changed quickly if the new, target concentration is a high value. Suppose that it was necessary to increase the concentration of a stable, DNA-binding protein by twofold in order to initiate a new, developmental program. Increasing transcription tenfold would cause the concentration to increase faster than would a twofold increase in synthesis. This is akin to the principle of treating patients with an initial, high concentration of a stable drug, or "loading dose," in order to achieve a therapeutic effect. It is possible that the phenomenon of enzyme overshoot, such as occurs for some hepatic enzymes at the time of birth, results from such a requirement for a rapid change. Of course, regulation by translational control or covalent modification circumvents this problem for mRNAs and proteins that are produced constitutively.

Fig. 5.5. The gene for tyrosine aminotransferase (TyrAT) includes extra segments at both ends that cause the enzyme to have a short half-life in vivo, whereas the gene for aspartate aminotransferase (AspAT) lacks these segments and is stable in vivo.

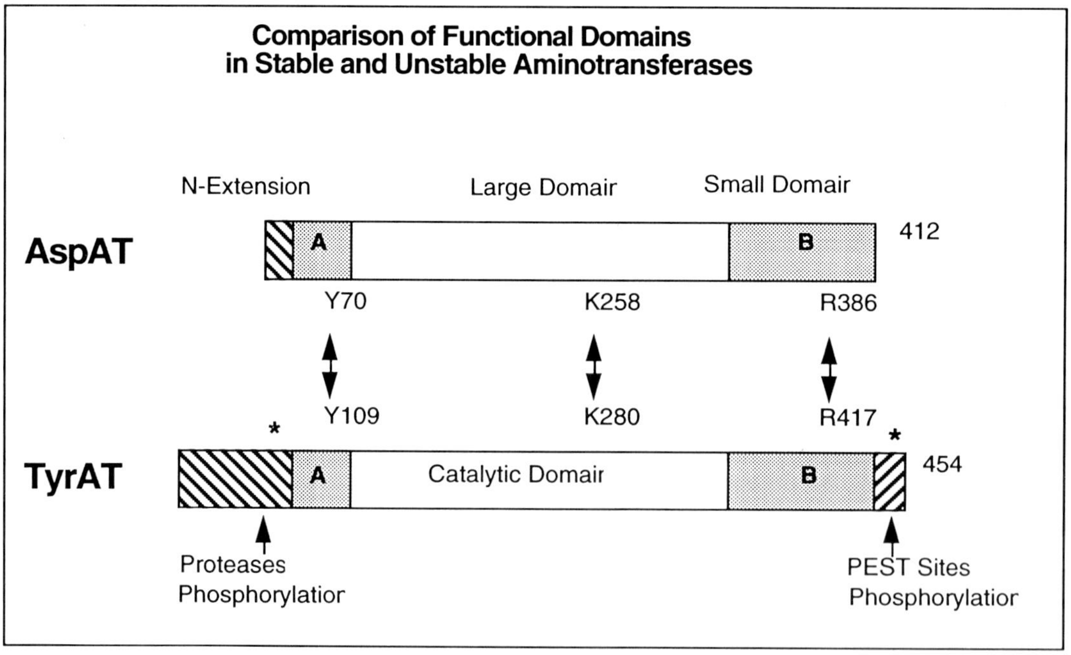

REFERENCES

1. Goldstein JL, Brown MS. Regulation of the mevalonate pathway. Nature 1990; 343:425-430.
2. Parker RA, Miller SJ, Gibson DM. Phosphorylation of native 97 kDa 3-hydroxy-3-methylglutaryl-coenzyme A reductase from rat liver. Impact on activity and degradation of the enzyme. J Biol Chem 1989; 254:4877-4887.
3. Pilkis SJ El-Maghrabi MR, Claus TH. Hormonal regulation of hepatic gluconeogenesis and glycolysis. Annu Rev Biochem 1988; 57:755-784.
4. DeCrombrugghe B, Busby S, Buc H. Cyclic AMP receptor protein: role in transcription activation. Science 1984; 224:831-838.
5. Beale EG, Hartley JL, Granner D. N^6,$O^{2'}$-dibutyryl cyclic AMP and glucose regulate the amount of messenger RNA coding for hepatic phosphoenolpyruvate carboxykinase (GTP). J Biol Chem 1982; 257, 2022-2028.
6. Chrapkiewicz NB, Beale EG, Granner DK. Induction of the messenger ribonucleic acid coding for phosphoenolpyruvate carboxykinase in H2-II-E cells. J Biol Chem 1982; 257, 14428-14432.
7. Sasaki K, Cripe R, Koch S, Andreone TL, Petersen DD, Beale EG. and Granner DK. Multihormonal regulation of phosphoenolpyruvate carboxykinase gene transcription: the dominant role of insulin. J Biol Chem 1984; 259, 15242-15251.
8. Sehgal P, Derman E, Molloy GR, Tamm I, Darnell JE. 5,6-Dichloro-1-β-D-ribofuranosylbenzimidazole inhibits initiation of nuclear heterogeneous RNA chains in HeLa cells. Science 1976; 194:431-433.
9. Peavy DE. Taylor JM, Jefferson LS. Time courses of changes in albumin synthesis and mRNA in diabetic and insulin-treated diabetic rats. Amer J Physiol 1985; 248: E656-E663.
10. Hopgood MF, Ballard FJ, Reshef L, Hanson R. Synthesis and degradation of phosphoenolpyruvate carboxylase in rat liver and adipose tissue. Changes during a starvation-refeeding cycle. Biochem J 1973; 134: 445-453.
11. Lee K-L, Reel JR, Kenney FT. Regulation of tyrosine α-ketoglutarate transaminase in rat liver. IX. Studies on the mechanism of hormonal inductions in cultured hepatoma cells. J Biol Chem 1970; 245: 5806-5812.
12. Yagil G. Quantitative aspects of protein induction. Curr Top Cell Reg 1975; 9:183-286.
13. Towle HC, Mariash CN, Schwartz HL, Oppenheimer JH. Changes in the hepatic levels of messenger ribonucleic acid for malic enzyme during the induction by thyroid hormone and diet. Biochemistry 1981; 20: 3486-3492.
14. Hargrove JL, Schmidt FH. The role of mRNA and protein stability in gene expression. FASEB J 1989; 3: 2360-2370.
15. Hargrove JL, Hulsey M, Schmidt F, Beale E. A computer program for modeling the kinetics of gene expression. BioTechniques 1990; 8: 654-659.
16. Diesterhaft M, Noguchi T, Granner D. Regulation of rat-liver tyrosine-aminotransferase mRNA by hydrocortisone and by N^6, $O^{2'}$-dibutyryl-adenosine 3', 5'-phosphate. Eur J Biochem 1980; 108:357-365.

17. Olson PS, Thompson EB, Granner, DK. Regulation of hepatoma tissue culture cell tyroxine aminotransferase messenger ribonucleic acid by dexamethasone. Biochemistry 1980; 19: 1705-1711.
18. Rudack D, Chisholm EM, Holten D. Rat liver glucose 6-phosphate dehydrogenase. J Biol Chem 1971; 246: 1249-1254.
19. Winberry L, Holten D. Rat liver glucose-6-P dehydrogenase. J Biol Chem 1977; 252: 7796-7801.
20. Prostko C, Fritz RS, Kletzien RF. Nutritional regulation of hepatic glucose-6-phosphate dehydrogenase. Biochem J 1989; 258: 295-299.
21. Dozin B, Rall JE, Nikodem VM. Tissue-specific control of rat malic enzyme activity and messenger RNA levels by a high carbohydrate diet. Proc Natl Acad Sci USA 1986; 83, 4705-4709.
22. Goldman MJ, Back DW, Goodridge AG. Nutritional regulation of the synthesis and degradation of malic enzyme messenger RNA in duck liver. J Biol Chem 1985; 260: 4404-4408.
23. Felder MR, Watson G, Huff M, Ceci JD. Mechanism of induction of mouse kidney alcohol dehydrogenase by androgen. J Biol Chem 1988; 263: 14531-14537.
24. Fort P, Rech J, Vie A, Piechaczyk M, Bonnieu A, Jeanteur P, Blanchard J-M. Regulation of c-fos gene expression in hamster fibroblasts: initiation and elongation of transcription and mRNA degradation. Nucl Ac Res 1987; 15:5657-5667.
25. Van Beveren C, van Straaten F, Curran T, Muller R, Verma IM. Analysis of FBJ-MuSV provirus and c-fos (mouse) gene reveals that viral and clelular fos gene peoducts have different coarboxy termini. Cell 1983; 32:1241-1255.
26. Rogers S, Wells R, Rechsteiner M. Amino acid sequences common to rapidly degraded proteins: the PEST hypothesis. Science 1986; 234:364-368.
27. Shaw G, Kamen R. A conserved AU sequence from the 3' untranslated region of GM-CSF mRNA mediates selective mRNA degradation. Cell 1986; 46:659-667.
28. Wreschner DH, Rechavi G. Differential mRNA stability ot reticulocyte ribonucleases correlates with 3'-noncoding $(U)_nA$ sequences. Eur J Biochem 1988; 172:333-340.
29. Segal HL, Kim YS. Glucocorticoid stimulation of the biosynthesis of glutamic-alanine transaminase. Proc Natl Acad Sci USA 1963; 50: 912-915.
30. Perry ST, Lee KL, Kenney FT. Reconstitution of aminotransferases in vivo and their metabolic turnover. Arch Biochem 1979; 195: 362-367.
31. Stiles CD, Lee K-L, Kenney FT. Differential degradation of messenger RNAs in mammalian cells. Proc Nat Acad Sci USA 1976; 73:2634-2638.
32. Pave-Preux M, Ferry N, Bouguet J, Hanoune J, Barouki R. Nucleotide sequence and glucocorticoid regulation of the mRNAs for the isoenzymes of rat aspartate aminotransferase J Biol Chem 1988; 263:17459-17466.
33. Jansonius JN, Eichele G, Ford GC, Picot D, Thaller C, Vincent M. Spatial and covalent structure of mitochondrial aspartate aminotransferase. In: Christen P, Metzler D, eds. Transaminases . New York: John Wiley and Sons, 1985.

34. Hargrove JL, Scoble HA, Mathews WR, Baumstark BR, Biemann K. The structure of tyrosine aminotransferase. Evidence for domains involved in catalysis and enzyme turnover. J Biol Chem 1989; 264: 45-63.
35. Mehta PK, Hale TI, Christen P. Evolutionary relationships among aminotransferases, Eur J Biochem 1989; 186:249-253.
36. Stellwagen RH. Involvement of sequences near both amino and carboxyl termini in the rapid intracellular degradation of tyrosine aminotransferase. J Biol Chem 1992; 267:23713-23721.

CHAPTER 6

Physiological Modeling of Gene Expression

Between the characters that are used by the geneticist and the genes that his theory postulates lies the whole field of embryonic development, where the properties implicit in the genes become explicit in the protoplasm of the cells. Here we appear to approach a physiological problem, but one that is new and strange to the classical physiology of the schools.

—Thomas Hunt Morgan

MODELING THE CONSEQUENCES OF GENE EXPRESSION

This volume is an introduction to modeling gene expression, and was conceived with the notion of explaining how to use simulation software to model the central elements of gene expression. Because most kinetic models that have historically been used in molecular biology are examples of compartmental analysis, it was useful to point out the connections between the early kinetic models and this form of systems analysis, especially because several powerful microcomputer programs employ this technique. However, as depicted in Figure 1.1, the central mystery of genetics has always been the way in which the genes give rise to the phenotype, whether this meant a physical trait such as eye or coat color, or a step in a metabolic pathway.

At its highest level, mathematical modeling is used not just to analyze quantities and rates, but also to interconnect different aspects of medically or economically important conditions, and to make predictions concerning the causes of chronic diseases, the spread of epidemics, or the impact of a new strategy for disease prevention. There is an ample literature devoted to topics of this kind, and this chapter will provide a few examples. The range of these models runs from the basic ones presented in the prior chapters to highly technical approaches that require advanced mathematical skills to grasp.[1]

MECHANISTIC MODELS OF BACTERIAL METABOLISM

Bailey and colleagues have published a series of detailed models that describe the major metabolic features of *Escherichia coli*, including the replication of chromosomal and plasmid DNA, general transcription and translation of bacterial and plasmid mRNA, and specific details concerning the *lac* operon.[2-6] The *lac* model simulated the probability of binding for repressors and activators of *lac* transcription, using the binding affinities of each protein and a measure of transcriptional efficiency.[2,3] One conclusion from the work was that plasmids interfere with transcription from the *lac* operon. Subsequently, a larger scale model was produced that simulated DNA polymerase activity, the precursor pool of deoxyribonucleotides, RNA polymerase and the ribonucleotides, the ribosomes and the amino acids. The assumptions used with respect to the mRNA and protein compartments were fundamentally the same as described in prior chapters. Both small-scale and large-scale models were developed, following precedents with as few as four compartments, but eventually producing a model with over 30 variables. The model developed by Peretti and Bailey[4,5] was designed to test the effect of parasitic interactions of plasmids carried at various copy numbers on the cellular growth rate. An interesting prediction of the model was that an increased plasmid copy number would eventually decrease the stability of the cloned mRNA. The interested reader is referred to this work for an excellent example of the power of kinetic modeling to offer insights concerning mechanisms of system behavior, and the necessity of comparing theoretical results with experimental data.

STEROL-DEPENDENT REGULATION OF HMG-CoA REDUCTASE GENE EXPRESSION

Education should be an important use of kinetic modeling, which permits students to deal with data from the clinical literature in medical genetics. Chronic diseases such as coronary vascular disease, obesity, and diabetes have metabolic components that affect the expression of specific genes, and altered gene expression regulates enzymes that modify and sometimes hinder the outcome of treatment.[7,8] A classic example of a condition that involves altered gene expression is hypercholesterolemia. The regulation of cholesterol metabolism is impaired in individuals who either have an inherited disorder of lipoprotein metabolism, or whose diets and lifestyles include risk factors such as high intake of saturated fat and cholesterol and failure to exercise regularly or to quit smoking cigarettes.[9,10] Dietary modification is recommended as one means of treating these chronic diseases, along with administration of drugs that reduce the absorption of cholesterol from the gut, or that inhibit enzymes of cholesterol metabolism. Among the drugs that lower cholesterol levels is a class of inhibitors that block the enzyme, 3-hydroxy-3-methylglutaryl coenzyme A reductase (HMG CoA reductase), which controls the rate of cholesterol biosynthesis under many circumstances. Typical inhibitors of HMG CoA reductase are lovastatin, compactin, and derivatives of these agents.[11]

The goal of simulating the effects of diet and medications on cholesterol synthesis was chosen because the genes involved in this network affect cardiovascular disease, which is the major cause of death in the U.S. population. Also, there is a great deal of information about the regulation of the genes, enzymes, and lipoproteins that are involved in cholesterol metabolism. Another reason for interest is that the transcriptional and post-transcriptional effects of oxidized metabolites of cholesterol (the oxysterols) represent one of the first nutrient-gene feedback systems that has been characterized in mammals.[7] Recently, nuclear factors that bind to the sterol regulatory elements in DNA have been identified,[12] so it is likely that the mechanism of this negative feedback system will soon be understood at a molecular level.

THE CONTROL OF CHOLESTEROL SYNTHESIS AND ELIMINATION

The pathway for cholesterol biosynthesis and its intracellular controlling systems are depicted in Figure 6.1. Cholesterol enters the blood-

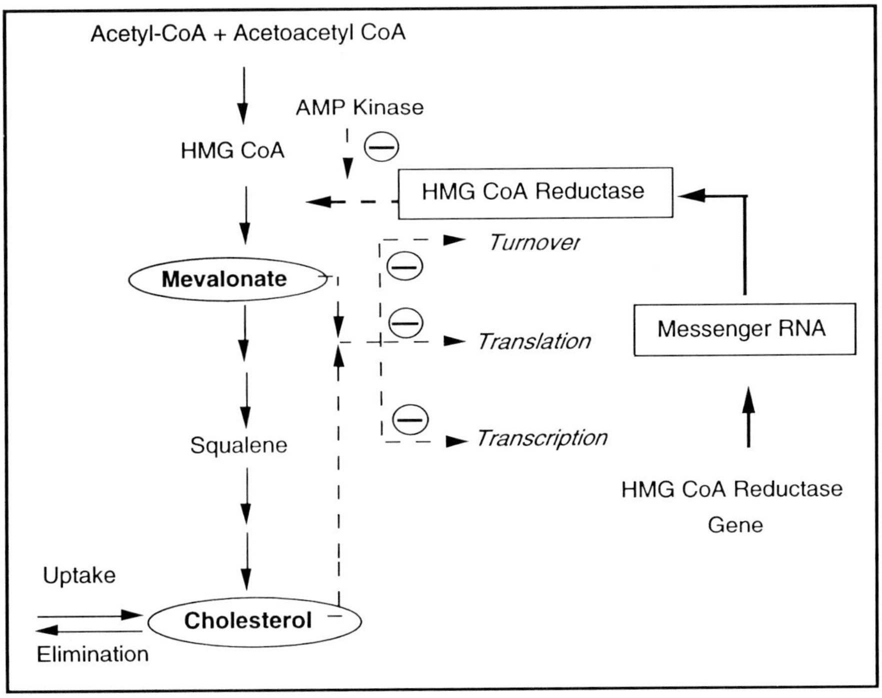

Fig. 6.1. The pathway for cholesterol biosynthesis is shown with the principle feedback controls on the activity of HMG CoA reductase. HMG CoA reductase controls the conversion of HMG Coenzyme A to mevalonic acid, which is converted by several enzymatic steps to squalene and then to cholesterol. Mevalonate and cholesterol both increase the rate of degradation of HMG CoA reductase, and this causes the concentration of the enzyme to decline fivefold. Metabolites of mevalonate also inhibit the translation of the mRNA, thereby reducing the rate of enzyme synthesis fivefold. Elevated cholesterol inhibits transcription of the gene by up to eightfold. These negative feedback loops produce a dynamic range of 200-fold for the concentration of HMG CoA reductase, and further negative control is exerted by inhibition due to phosphorylation by an AMP-dependent protein kinase.

stream not only by way of chylomicrons that are formed in the intestinal tract, but also as very low density lipoprotein (VLDL) particles, which are precursors to LDL-cholesterol. The liver and small intestines are responsible for the majority of VLDL synthesis. Although cholesterol synthesis requires several enzymes, under most conditions HMG CoA reductase controls the net rate of production.[13] During the course of a day, HMG CoA reductase activity in liver undergoes a pronounced cycle, with the minimum values occurring during the period of fasting, and the maximum after meals have been taken; the half-time for this diurnal cycle is about 3 hours.[13,14] This enzyme is active in all tissues because the product, mevalonic acid, is the precursor not only for sterols but also for isoprenoids that are essential for the formation of modified nucleosides, dolichol, and other compounds that are required to complete cell division. However, cholesterol that is exported from the liver is packaged along with triglycerides into lipoproteins that make up the VLDL particle.

In addition to its role in cholesterol biosynthesis, the liver is the major route for eliminating cholesterol from the body. In this process, LDL- and HDL-cholesterol is taken in from the blood plasma, and is either stored in esterified form, or is converted to bile acids by specific enzymes. The bile acids are secreted into the gall bladder and intestinal tract, and the main path of cholesterol elimination takes place via the feces. Each of these pathways is regulated by sterols; therefore, we have chosen to model the role of liver parenchymal cells in cholesterol metabolism.

Mevalonic acid exerts negative control over HMG CoA reductase by acting at two distinct levels. Mevalonate or a related metabolite decreases the rate of translation of the mRNA that encodes HMG CoA reductase by up to fivefold;[15] it also increases the rate of degradation of HMG CoA reductase itself.[16] The negative feedback exerted by mevalonate augments that produced by sterols or their metabolites. Incubation of cells with LDL cholesterol or oxysterols inhibits the transcription of the gene for HMG CoA reductase by up to eightfold,[17] and also accelerates the rate of degradation of HMG CoA reductase.[18,19] These two effects both cause the level of the enzyme to decline, which helps protect cells against accumulating toxic levels of these compounds. However, when an inhibitor of HMG CoA reductase is administered, the concentration of mevalonate and cholesterol declines, and the net effect is to increase the concentration of the enzyme as much as 200-fold.[7] As long as the inhibitor is present, the enzymatic activity is low, but there is concern that a patient who stopped taking the medication could experience a rebound synthesis of excessive sterol because the half-life of this class of inhibitors is less than 2 hours in human patients (Pan, 1991).[20]

OXYSTEROL EFFECTS ON HMG CoA REDUCTASE

The purpose of the model described here is to simulate the feedback control that is exerted by sterols and inhibitors of HMG CoA reductase on the transcription, translation, and metabolic stability of HMG CoA reductase and its mRNA. In order to focus on the sterol-dependent system, other levels of regulation such as altered thyroid status, phosphorylation by the AMP-dependent protein kinase, and altered bile acid synthesis were not included. For this purpose, a computer program was created using the STELLA®II tool for mathematical modeling according to the principles we have explained earlier. This tool enables the user to create a diagram that includes the major points of regulation for a system of interest; as the diagram is set up, initial values for concentrations of the components are entered into dialog boxes, and quantitative relationships or equations that describe the kinetic behavior are similarly entered. STELLA®II constructs a computer program from the diagram by setting up finite difference equations that predict the concentrations of material in the indicated compartments, and the movement of material from one space to another. This output is entered into tables or graphs in the manner specified by the operator.

Figure 6.2 depicts a two-compartmental model for production of HMG CoA reductase and its mRNA. The diagram includes the major points of regulation by the pool of sterol within the hepatocyte, as well as the points affected by inhibitors of HMG CoA reductase. The compartment on the upper left of the figure represents the concentration of the mRNA that encodes this enzyme. The mRNA accumulates until the net production is balanced by the rate of mRNA decay; degradation is thought to take place at a rate proportional to concentration (first order rate constant). These features are indicated by the flows into and out of the mRNA compartment (flows are designated by arrows with circular control elements). In turn, the mRNA is translated to generate the active protein, which is represented by the box on the right side of Figure. 6.2; the protein is also degraded at a rate that is proportional to concentration. These processes of biosynthesis and degradation combine to generate a predictable concentration of enzyme, and the time required for the concentration to change depends on the half-lives of the mRNA and the protein. In the case of HMG CoA reductase, the rates of synthesis and decay are not fixed; each point is regulated by the cellular concentrations of cholesterol and the product, mevalonate, as well as by hormones from the endocrine system.

A model of cholesterol metabolism must link the quantity of the enzyme to its capacity to synthesize cholesterol. This feature is indicated by the arrow to the converter labeled "Reductase activity." For the present model, the enzyme was assumed to follow Michaelis-Menten kinetics, and this formula was entered into the dialog box associated with "Reductase activity." The enzyme controlled the rate of cholesterol synthesis, as indicated by the arrow connecting "Reductase activity" to "Sterol synthesis." To simplify the model, it was assumed that the precursor supply did not change. It would be feasible to permit substrate to vary and employ formulae derived from metabolic control theory to demonstrate how enzyme activity acts on substrate to control flux through the pathway.[21]

The main point of regulation of the mRNA concentration is transcriptional initiation, which is increased by thyroid hormone and decreased by metabolites of cholesterol and mevalonic acid. In the model depicted here, transcription takes place at a fixed rate unless the cholesterol concentration changes. As the concentration increases, the rate of transcription decreases from the initial value. The decline in transcription was assumed to follow a sigmoidal response curve that is predicted from a form of the Hill equation:

$$T = T_{max}\ (1 - C^n/[C^n + (C_{0.5})^n]) \qquad \text{Eq. 6.1}$$

where T_{max} equals the maximum rate of transcription, C is the concentration of inhibitory cholesterol (LDL cholesterol or oxysterol), and $C_{0.5}$ is the concentration of cholesterol that gives half-maximal repression of the rate of transcription.[22] A value of 5 μmol/gram was chosen for $C_{0.5}$. The exponent, n, is an apparent Hill coefficient that was

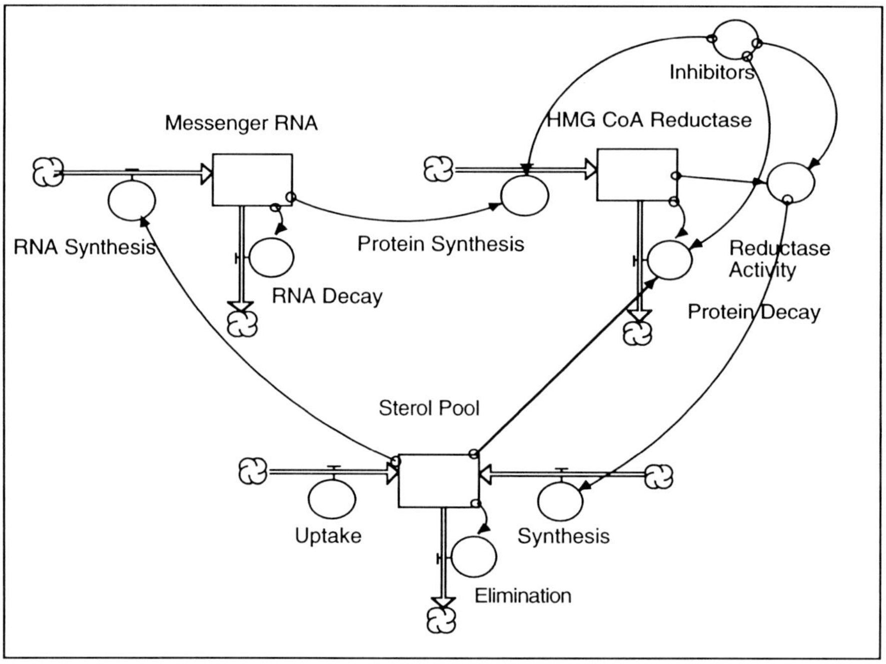

Fig. 6.2. A computer program was designed to simulate the control of HMG CoA reductase in cholesterol metabolism using STELLA®II. The mRNA accumulates in a pool represented by the rectangle (upper left) as a function of its rates of synthesis and decay. The mRNA is translated to cause a proportional increase in synthesis of HMG CoA reductase, as indicated by the arrow that connects the mRNA pool to the controller for protein synthesis. HMG CoA reductase then accumulates as a function of its rates of synthesis and decay. This enzyme initiates the synthesis of cholesterol at a rate that is proportional to the reductase activity (circle on right side). As the sterol accumulates, it causes the rate of RNA synthesis to decrease and the rate of protein turnover to increase (connecting arrows from sterol pool, bottom of diagram). When a medication such as lovastatin is given, HMG CoA reductase is inhibited and production of mevalonate ceases. This prevents the negative feedback on mRNA translation and protein turnover (arrows from the controller labeled Inhibitors in upper right). As a result, the inhibitors cause an increase of (inactive) HMG CoA reductase of up to 200-fold.

assumed to equal 2, since many DNA-binding proteins are dimers. This provides a sharp inflection, in contrast to the flattened response when the exponent is 1, as is typical of Michaelis-Menten sensitivity. All the sterol-regulated functions were assumed to respond with the same sensitivity.

The rates of synthesis and activity levels reported for hepatic HMG CoA reductase and cholesterol in rats were chosen for the initial modeling. Reference patterns for responses of animals to various diets and inhibitors were obtained from the literature; the ranges and values used in the simulations are shown in Tables 6.1 and 6.2.

RESULTS OF COMPUTER-BASED SIMULATIONS

The intent of this project was to reproduce the response of the sterol-dependent feedback system that is expected to result from altered synthesis, uptake, or elimination of intracellular sterols, as well as to show how treatment with inhibitors of HMG CoA reductase would affect these processes. The predicted consequences of treatment with an inhibitor such as lovastatin or mevinolin are shown in Figure. 6.3, under the assumption that complete inhibition of HMG CoA reductase is achieved. As the activity approaches zero, synthesis of mevalonate and cholesterol is prevented, and the sterol pool begins to decline. As the sterol pool declines to very low levels, the sterol- and mevalonate-dependent inhibition of the gene for HMG CoA reductase is relieved. This leads to an eightfold increase in transcription, and also causes the enzyme to decay at a slower rate. The combined effects on transcription, mRNA translation, and protein stabilization cause a 200-fold increase in enzyme concentration, but the enzyme will be catalytically inactive unless the inhibitor is removed.

To demonstrate the action of the negative feedback system, a simulation was performed in which the rate of cholesterol uptake was increased fourfold. The model suggests that this treatment would produce an increase in intracellular sterol pools that is smaller than the magnitude of change for total uptake. The reason for this prediction is that the increased sterol down-regulates the HMG CoA reductase gene at transcriptional and post-transcriptional levels, and this causes biosynthesis of sterol to decrease (Fig. 6.4). Similarly, reduced dietary intake of cholesterol would increase the activity of HMG CoA reductase and cholesterol synthesis; this causes a smaller degree of change than would occur in the absence of the negative feedback system.

Suppose that a patient who was being treated with the medication neglected to continue with the therapeutic dosage; it is reasonable to ask what the consequences would be for cholesterol synthesis. The answer depends on the rate at which the drug is cleared from the body, relative to the rate at which the enzyme is degraded. The half-time for decay of the enzyme is expected to be about 3 h under usual conditions, and when the enzyme is reactivated, it would begin to synthesize cholesterol and mevalonic acid so that the usual negative feedback would be reestablished. If the enzyme was truly induced by 200-fold, it would be expected that about 6-7 half-times would elapse before the enzyme approached its normal range; therefore, elevated synthesis would continue for about 18-21 hours. Because the half-life for inhibitors of HMG CoA reductase in the blood plasma is shorter than 3 hours,[20] the possibility that the enzyme activity might rebound deserves to be tested.

DIOXIN METABOLISM, GENE EXPRESSION, AND RISK ASSESSMENT

In an effort to gain insight into the reason that dioxin produces hepatic cancer in female rats but not in males, Kohn and colleagues

Table 6.1. Kinetic data used for modeling HMG CoA reductase gene expression[a]

Molecular weight:	97,092 daltons (887 amino acids)	
Specific Activity:	5.2 μmoles/min/mg purified enzyme	
Activity in liver:	2.5 μmoles/min in 100 grams of rat liver	
Concentration in liver:	ca. 50 nanomoles of enzyme/gram	
K_M for HMG CoA:	1 μM	
K_M for NADPH:	40 μM	
Range of activity in liver microsomes:		
Normal	420 pmol/min/mg protein	
Cholesterol	14 pmol/min/ mg protein	
Cholesterol + mevalonolactone	2.5 pmol/min/mg protein	
Enzyme half-life	*(Rate of enzyme decay)*	
No additions	2.25 h	($0.308\ h^{-1}$)
plus sterols	1.25 h	($0.554\ h^{-1}$)
plus compactin	5 h	($0.139\ h^{-1}$)
Size of mRNA	4.8 kilobases x 321 = 1.54×10^6 daltons	
Half-life for mRNA	3 h	
Rate of mRNA decay	$0.231\ h^{-1}$	
Rate of transcription	24-922 ppm	

[a]Values were calculated from data obtained in the references.[7-25]

Table 6.2. Kinetic values for cholesterol in rat liver[a]

Cholesterol concentration	5.7 μmol $gram^{-1}$
Rate of cholesterol synthesis	2.7 μmol $gram^{-1}\ h^{-1}$
Absorption from diet (chylomicrons)	0.7-1.5 μmol $gram^{-1}\ h^{-1}$
Secretion into bile	0.1 μmol $gram^{-1}\ h^{-1}$
Uptake from blood plasma (LDL particles)	0.6 μmol $gram^{-1}\ h^{-1}$
Secretion into blood plasma	$0.7\ h^{-1}$

[a]Values were calculated from data obtained in the references.[25] These figures are approximate because diet and treatments vary among different reports, and because rates of synthesis, uptake and elimination are each subject to regulation.

recently generated a very interesting mechanistic model of the effects of dioxin (2,3,7,8-tetracholodibenzo-dioxin) on gene expression in liver.[26-30] The model is noteworthy for its completeness. Using the SCoP simulation tool, equations were written for dioxin's tissue distribution, clearance rates, uptake into liver, and effects on cytochrome P450 isoenzymes. The model included modulation of estradiol metabolism by cytochrome P450 IA2, and a feedback effect through tumor necrosis factor-α on the epidermal growth factor receptor, which was related to cellular proliferation. The model was capable of reproducing the effects of dioxin on two isoenzymes of cytochrome P450 under steady state conditions, and predicted effects on the aryl hydrocarbon receptor, estrogen, and growth factor receptors. This is an excellent example of the ability of systems models to integrate a information from several fields, and to produce useful predictions about consequences for specific diseases.

In addition to the usefulness of computer simulation in research, its potential as a teaching tool in graduate education should be indicated. All medical issues are highly complex, and the process of designing a predictive model is a useful exercise that can stimulate discussion and enhance awareness of issues. The work on dioxin is a case

Fig. 6.3. The predicted response of HMG CoA reductase and intracellular cholesterol to treatment with a reductase inhibitor is depicted. When a reductase inhibitor such as lovastatin is added to cultured cells, the sterol production is inhibited and a new level is achieved that depends on the balance between uptake and elimination. Because the inhibitor interferes with the negative feedback on enzyme synthesis, HMG CoA reductase accumulates to a very high level.

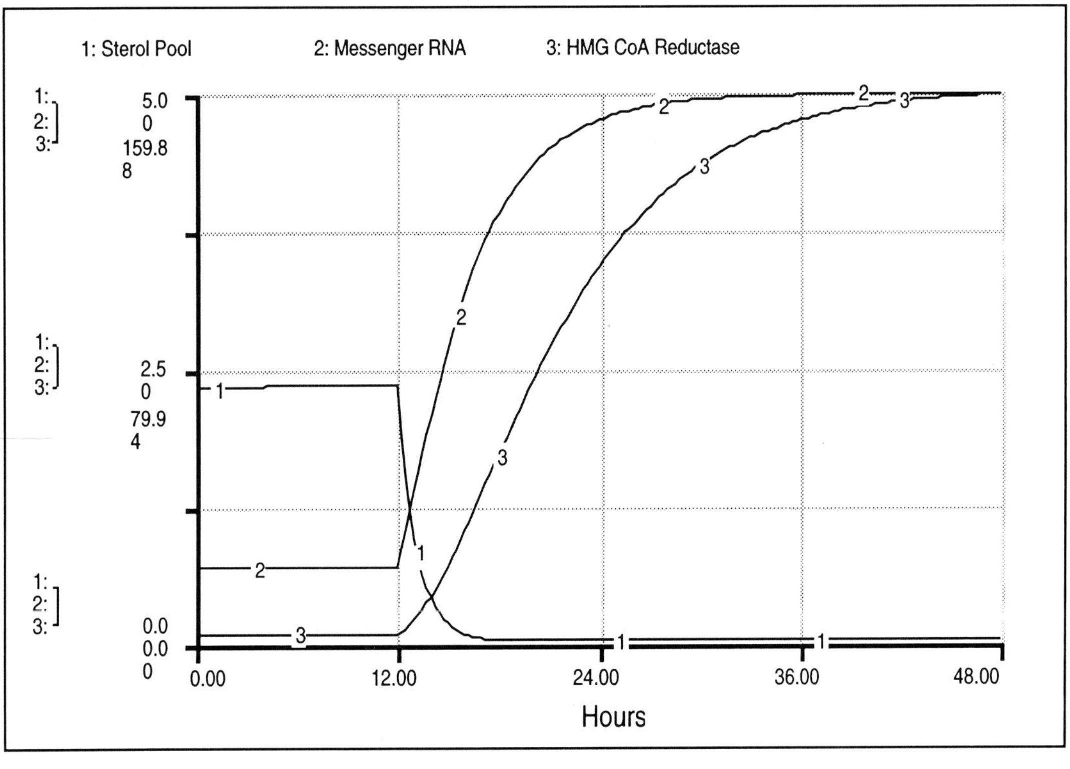

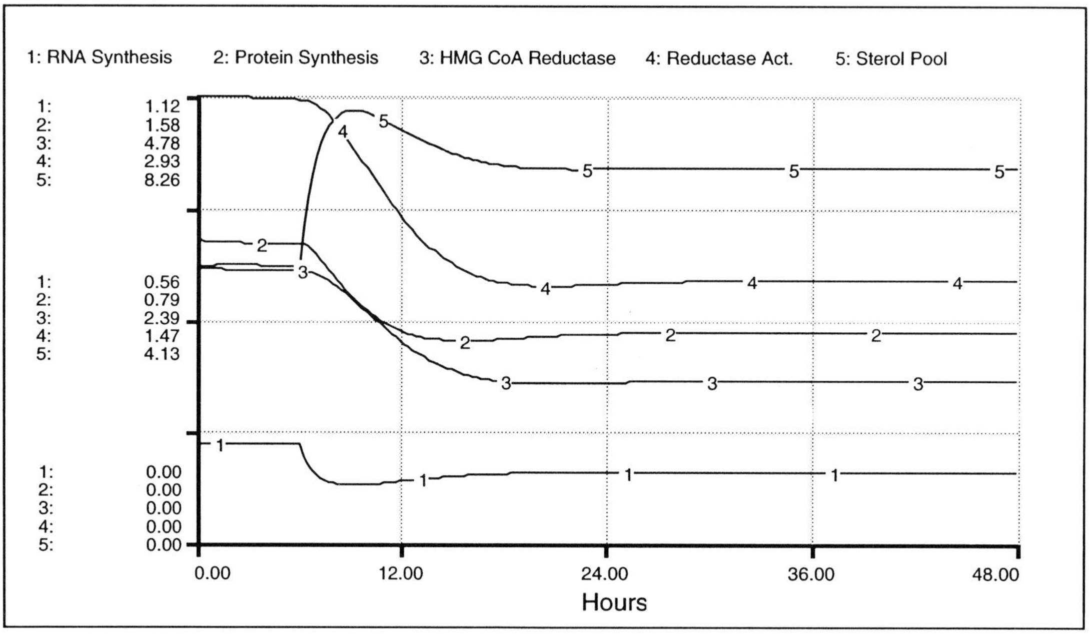

Fig. 6.4. Demonstration of the negative feedback system that regulates HMG CoA reductase and intracellular cholesterol. When cholesterol intake was increased fourfold at 6 hours, the sterol pool rose from 5 to 8 μmol/g of tissue (line 5). The increased sterol caused the rate of transcription to decrease (line 1) and also caused the rate of degradation rate of HMG CoA reductase to increase. As a result, the enzyme activity declined and reduced the rate of cholesterol synthesis (line 4). Therefore, the fourfold increase in cholesterol absorption caused only a 1.5-fold increase in the intracellular sterol pool, demonstrating the effectiveness of the negative feedback system.

in point, which has implications for policy making as well as for testing of hypotheses. The increasing complexities of research and education call for an integrative tool, and computer-assisted simulation has much potential for enhancing both areas.

References

1. Bellman K. Molecular Genetic Information Systems. Modelling and Simulation. Berlin: Akademie Verlag, 1983.
2. Lee SB, Bailey JE. Analysis of growth rate effects on productivity of recombinant *Escherichia coli* populations using molecular mechanistic models. Biotechnol Bioeng 1984; 26:66-73.
3. Lee SB, Bailey JE. Genetically structured models for *lac* promoter-operator function in the *Escherichia coli* chromosome and in multicopy plasmids: *lac* operator function. Biotechnol Bioeng 1984; 26:1372-1382.
4. Peretti SW, Bailey JE. Mechanistically detailed model of cellular metabolism for glucose limited growth of *Escherichia coli* B/r-A. Biotechnol Bioeng 1986; 28: 1672-1689.
5. Peretti SW, Bailey JE, Lee JJ. Transcription from plasmid genes, macromolecular stability, and cell-specific productivity in *Escherichia coli* carry-

ing copy number mutant plasmids. Biotechnol Bioeng 1989; 34:902-908.
6. Bailey JE. Host-vector interactions in *Escherichia coli.* Adv Biochem Eng 1993; 48:30-52.
7. Goldstein JL, Brown MS. Regulation of the mevalonate pathway. Nature 1990; 343: 425-430.
8. Motulsky AG et al. Diet and Health. Implications for Reducing Chronic Disease Risk, Washington DC.: National Academy Press, 1989.
9. Gotto A, Pearson TA, Felder D, Stoy D. Coronary Heart Disease and the Lipid Triad. A Core Curriculum for Physicians, Cedar Grove NJ: HealthScan, Inc, 1992.
10. Hwa JJ, Zollman S, Warden CC et al. Genetic and dietary interactions in the regulation of HMG CoA reductase gene expression. J Lipid Res 1992; 33: 711-725.
11. Myant NB. In: Cholesterol Metabolism, LDL, and the LDL Receptor, San Diego: Academic Press, 1990.
12. Stark HC, Weinberger O, Weinberger J. Common double and single-stranded DNA binding factor for a sterol regulatory element. Proc Natl Acad Sci USA 1992; 89: 2180-2184.
13. Beg ZH, Brewer HB. Regulation of the liver 3-hydroxy-3-methylglutaryl CoA reductase. Curr Top Cell Reg 1981; 20: 139-184.
14. Higgins M, Kawachi T, Rudney H. The mechanism of the diurnal variation of hepatic HMG CoA reductase activity in the rat. Biochem. Biophys. Res Commun 1971; 45: 138-144.
15. Nakanishi M, Goldstein JL, Brown MS. Multivalent control of 3-hydroxy-3-methylglutaryl coenzyme A reductase. Mevalonate-derived product inhibits translation of mRNA an accelerates degradation of enzyme. J Biol Chem 1988; 263: 8929-8937.
16. Edwards PA, Lan SF, Fogelman AM. Alterations in the rates of synthesis and degradation of rat liver 3-hydroxy-3-methylglutaryl coenzyme A reductase produced by cholestyramine and mevinolin. J Biol Chem 1983; 258: 10219-10222.
17. Clarke C, Fogelman A, Edwards P. Transcriptional regulation of the 3-hydroxy-3-methylglutaryl coenzyme A reductase gene in rat liver. J Biol Chem 1985; 260: 14363-14357.
18. Chin DJ, Gil G, Faust JR, Goldstein JHL, Brown MS, Luskey KL. Sterols accelerate degradation of hamster 3-hydroxy-3-methylglutaryl coenzyme A reductase encoded by a constitutively expressed cDNA. Mol Cell Biol 1985; 5: 634-641.
19. Ness GC, Keller RK, Pendleton LC. Feedback regulation of hepatic 3-hydroxy-3-methylglutaryl coenzyme A reductase activity by dietary cholesterol is not due to altered mRNA levels. J Biol Chem 1991; 266:14854-14857.
20. Pan HY. Clinical pharmacology of pravastatin a selective inhibitor of HMG CoA reductase. Eur J Clin Pharmacol 1991; 40 (Suppl. 1): S15-S18.
21. Cornish-Bowden A, Hofmeyer J-H. S. MetaModel: a program for modeling and control analysis of metabolic pathways on the IBM PC and compatibles, CABIOS 1991; 7: 89-93.
22. Ralston DM, O'Halloran TV. Ultrasensitivity and heavy metal selectivity of the allosterically modulated MerR transcription unit. Proc Natl Acad

Sci USA 1990; 87: 3846-3850.
23. Simonet WS, Ness GC. Turnover rates of HMG CoA reductase mRNA: role of pituitary and thyroid hormones. Biochem Biophys Res Commun 1987; 146: 1033-1039.
24. Chin DJ, Gil G. Russell DW, Liscum L, Luskey KL, Basu SK, Okayama H, Berg P, Goldstein JL, Brown MS. Nucleotide sequence of 3-hydroxy-3-methylglutaryl coenzyme A reductase a glycoprotein of the endoplasmic reticulum. Nature 1984; 308: 613-617.
25. Turley SD, Dietschy JM. The metabolism and excretion of cholesterol by the liver. In: Arias IM, Jakoby WB, Popper H, Schacter D, Shafritz DA, eds, The Liver. New York: Raven Press, 1989.
26. Kohn MC, Lucier GW, Clark GC, Sewall C, Tritscher AM, Portier CJ. A mechanistic model of effects of dioxin on gene expression in the rat liver. Toxicol Appl Pharmacol 1993; 120, 138-154.
27. Kohn MC, Melnick RL. Species differences in the production and clerarance of 13 butadiene metabolites: a mechanistic model indicates predominantly physiological not biochemical contrl. Carcinogenesis 1993; 14:619-628.
28. Kohn MC, Portier CJ. Effects of the mechanism of receptor mediated gene expression on the shape of the dose-response curve. Risk Analysis 1993; 13:565-572.
29. Kohn MC, Lucier GW, Clark GC, Sewall C, Tritscher AM, Portier CJ. A mechanistic model of effects of dioxin in the rat liver. Toxicol Appl Pharmacol 1993; 120: 138-154.
30. Portier C, Tritscher A, Kohn M, Sewall C, Clark G, Edler L, Hoel D, Cucier G. Ligand/receptor binding for 2,3,7,8-TCDD: Implications for risk assessment. Fund Appl Toxicol 1993; 20:48-56.

INDEX

Page numbers in italics denote figures (f) and tables (t).

The kinetic modeling techniques described in this book, *Kinetic Modeling of Gene Expression,* use the STELLA® II program for the Macintosh which can be purchased for $295 from:

High Performance Systems, Inc.
45 Lyme Road
Hanover, New Hampshire 03755
U.S.A.

Phone 603.643.9636
Fax 603.643.9502

A STELLA® II Run-time program designed by the author, James L. Hargrove, can accompany the text of the book or can be used as a stand-alone instructional tool for kinetic modeling of gene expression. The Run-time program simulates mammalian gene expression using the principles of compartmental analysis. It functions just like the original STELLA®II simulation program except that it does not permit the user to save changes to numerical values of model configuration. This module can be changed during use, but when the program is turned off the starting conditions will be restored to their default values.The goal of the model is to provide molecular biologists with a means to quantitatively simulate the consequences of transitional activation and post-translational controls, including altered translation or altered stability of mRNAs and proteins. The STELLA® II Run-time program for the Macintosh can be purchased for $39 from:

R.G. Landes Co.
P.O. Box 4858
Austin, Texas 78765
U.S.A.

Phone 512.863.7762
Fax 512.863.0081